Nanostructured Materials for Engineering Applications

Carlos Pérez Bergmann
and Mônica Jung de Andrade (Eds.)

Nanostructured Materials for Engineering Applications

Springer

Editors

Dr. Carlos Pérez Bergmann
Universidade Federal do
Rio Grande do Sul
Escola de Engenharia
Depto. Materiais
Av. Osvaldo Aranha 99-7 Andar
90035-190 Porto Alegre Rio
Grande do Sul
Brazil
Tel.: +555133083405

Dr. Mônica Jung de Andrade
Universidade Federal do
Rio Grande do Sul
Escola de Engenharia
Depto. Materiais
Av. Osvaldo Aranha 99-7 Andar
90035-190 Porto Alegre Rio
Grande do Sul
Brazil
E-mail:mja0612@gmail.com

ISBN 978-3-642-19130-5 ISBN 978-3-642-19131-2 (eBook)

DOI 10.1007/978-3-642-19131-2

Typesetting: Data supplied by the authors

Cover Design: Scientific Publishing Services Pvt. Ltd., Chennai, India

Printed on acid-free paper

9 8 7 6 5 4 3 2 1

springer.com

Preface

This book contains a brief overview of nanostructured materials for applications in Engineering, including some of its interfaced and supporting areas. After the introduction, it presents the general properties of nanomaterials, addressing special phenomena and potentials properties, without going into too much scientific detail of the physics and chemistry involved.

The next chapters present applications of nanostructured materials in the following fields: nanomagnetic materials, optoeletronic and ferroeletric applications, energy applications, bio-applications, catalysis, nanoreinforcements for nanocomposite materials, applications in refractory materials, adsorbent applications and use of natural and modified nanomaterials. For each application, it is described the last developments of laboratorial and/or industrial activities, as well as the trends in new materials and their technological processing.

We hope that the clear language and the application-oriented perspective are suitable for both engineers and students that want to discover the fascinating field of nanostructured materials.

31st of January 2011
Brazil

Porto Alegre
C.P. Bergmann
M. Jung de Andrade

Contents

1 Introduction

Carlos Pérez Bergmann

Universidade Federal do Rio Grande do Sul, Porto Alegre, Brazil
E-mail: bergmann@ufrgs.br

The present book entitled Nanostructured Materials for Engineering Applications is a part of the multidisciplinary series Engineering Materials.

Nanostructured Materials for Engineering Applications is composed of ten chapters that give a brief introduction to the different engineering applications of nanostructured materials. It addresses the latest developments in the field without going into too much scientific detail of the physics and chemistry involved, which will make the reading more interesting, motivating professionals and students in the field.

In the first chapter, the reader will learn about the phenomena involved with the properties of nanomaterials. In the following chapters nanostructured materials for different applications in Engineering are described, such as those used in Magnetic applications, Opto-electronic and Ferroelectric applications, Energy, Bio-applications, Catalysis, Nanoreinforcement for Nanocomposite materials, Refractory Materials, Adsorbent applications, as well as the use of Natural Modified Nanomaterials. The nanostructured materials used in the field and their processing, besides some practical examples of their use in laboratories and/or industry are described.

The clear language and the application-oriented perspective, in which the book is written, make it suitable for both engineers and students who want to discover the fascinating field of applications of nanostructured materials.

Some of the questions the reader might be asking himself are:

What are nanomaterials and nanostructured materials?

According to ISO TC 229: "Nanotechnology Standardization in the field of nanotechnology includes either or both of the following:

1. Understanding and control of matter and processes at the nanoscale, typically, but not exclusively, below 100 nanometers in one or more dimensions where the onset of size-dependent phenomena usually enables novel applications.

2. Utilizing the properties of nanoscale materials that differ from the properties of individual atoms, molecules, and bulk matter, to create improved materials, devices, and systems that exploit these new properties."

Thus, whenever one of its dimensions is below 100 nm, it is considered a nanomaterial, while whenever one of the constituents of the structure of the material has a dimension below 100 nm, it could be considered a nanostructured material.

What would be considered a Nanostructured Material for Engineering Applications?

As nanostructured materials for engineering applications we consider all structural and functional materials with specific engineering properties. Some important engineering properties are magnetism, strength, impact resistance, electrical and thermal conductivities, thermal expansion and high temperature resistance, among others. Consequently, this includes all nanostructured metals, polymers, ceramics and composites to create a range of products, from computer chips and aircraft wings to fuel cells and artificial organs.

To understand the correlation between the structure and properties of these materials of technological interest, one must "think small": understand these materials from their atomic structure / their crystal and cluster formation of atoms / their ions with organizational periodicity on a dimensional level unimaginably small, between 10^{-15} m and 10^{-10} m.

For applications in engineering, as a rule, one must consider the magnitude of the dimensional components normally used in devices and pieces of equipment or facilities for their technological use. Therefore, it is necessary to "think macroscopically", that is, to think about the macrostructure of the material for its properties of interest, stemming from the nanostructure, to obtain the desired function. In this context, the large dimensional distance between the nano- and macro-structure is established, which must be traversed for the successful use of nanostructured materials in engineering.

Is it safe to work with Nanostructured Materials for Engineering Applications?

Along this book, the reader will realize that the exponential growth of nanotechnology has enabled a great variety of "nanoproducts" to already reach the market. Engineering applications, transformation in chemical industries, medical products, agribusiness, electronics, defense and energy are some of the recipients of this technology.

However, it is important to outline that the regulations for nanomaterials are still in discussion and have not accompanied their advancement into the market. According to a recent study of the world stage in terms of regulation of "nanoproducts", U.S.A., Canada, Japan and the E.U. are more advanced [1]. Each new technology that is inserted in the market always needs a technical evaluation of risk/security that depends on the innovation of the technique or materials used. It is important to prevent possible risks during the different stages (production, use and degradation) of the product in order to guarantee the security of the human health and of the environment. Thus, technical know-how, weighing options and good sense are required whenever working/dealing with nanotechnology.

The potential offered by nanotechnology and its new and innovative applications are attracting increasing interest from various sectors, both academic and industrial. The removal of obstacles to the use of nanotechnology, enabling products based on nanostructured materials to be available soon to society, will take place as advances occur in the interaction between industry, research centers and the development of human resources. It is, therefore, essential to exchange information and experiences in the related areas. It is in this context that this book must be included.

Reference

[1] www.nanotech.law.asu.edu Accessed November 2010

2 Nanomaterials Properties

Antonio Gomes Souza Filho[1] and Solange Binotto Fagan[2]

[1] Departamento de Física, Universidade Federal do Ceará, P.O. Box 6030, 60455-900, Fortaleza-CE, Brazil
[2] Área de Ciências Tecnológicas, Centro Universitário Franciscano - UNIFRA, 97010-032, Santa Maria-RS, Brazil
E-mail: agsf@fisica.ufc.br, solange.fagan@gmail.com

Abstract. In this chapter we introduce some basic physical properties of nanomaterials compared with their bulk counterparts by using some particular nanostructured materials as model. The role of surface atoms and the quantum-induced effects in nanoscale materials are discussed. The notable electronic properties of nanosystems are also presented with emphasis given to the carbon nanotubes. We also evaluate the vibrational properties of nanomaterials and how these properties can be used for characterizing nanomaterials thus obtaining information about size, disorder, and morphology. The size-induced magnetic and ferroelectric properties of nanomaterials are also highlighted.

Keywords: nanomaterials, quantum effects, surface atoms, electronic properties, vibrational properties, magnetic properties.

2.1 Introduction

The new and remarkable properties of nanomaterials will influence the modern society in many aspects. Nanosized materials would not be so interesting and, both, nanoscience and nanotechnology would not be so attractive and exciting fields if nanosystems would behave like their bulk counterparts. The era of dealing with tiny objects has been gaining momentum in the past few years because of the industrial progress, the scientific ability to fabricate, model and manipulate things with a small numbers of atoms, and the almost daily discovery of novel size-induced phenomena.

The origin of the size-induced properties in nanomaterials depends basically on the surface phenomena (*extrinsic contribution*) and quantum confinement effects (*intrinsic contribution*). The surface to volume ratio increases rapidly when particle size decreases. A very simple model can be used for understanding this effect. Let us consider a spherical particle. The surface area (πD^2) to volume ($\pi D^3/6$)

ratio is readily evaluated as being proportional to D^{-1} (D is the nanoparticle diameter). The atoms at the surface have different properties because they present an asymmetric interaction, i.e., in one side they have atoms for interacting but these interactions are missing towards the external surface. The same happens with bulk materials but it is not relevant because in bulk the number of atoms at the surface is considerably smaller than that at the volume. This asymmetry is responsible for phenomena at the surface such as lattice distortion which is needed for accommodating the interacting forces and keeping the total energy as lowest as possible because the surface atoms are of higher energy than those of the bulk. This accommodation of the system has a consequence on its properties, i.e., they exhibit a gradient near the surface which is relevant up to a critical length l_c. The size of this critical length depends on the potential interaction between the constituents of the nanoparticles and determines two regimes for bulk and surface-based properties of the nanoparticles, i.e., for $D >> l_c$ ($D << l_c$) the properties of the nanoparticles are mainly determined by the bulk (surface) properties [1].

When a given material has, at least, one dimension being reduced to nanoscale the electronic wavevectors becomes quantized and the system exhibits discrete energy levels. The relevant length scale is the de Broglie wavelength, λ_{dB}, of an electron in one particular direction: if the dimension of the system in one dimension is lower than λ_{dB} we call it a two-dimensional (2D) system. Graphene, one atom thick layer exfoliated from graphite, is a very good example of a 2D system as well as semiconductor superlattices. One-dimensional (1D) system is obtained when two dimensions are lower than λ_{dB} values; nanowires and nanotubes belong to this category. Special is the case of the so-called quantum dots whose all three dimensions are lower than λ_{dB} and we name these as 0D systems. Their energy spectra are discrete and the system can be viewed as an artificial atom.

In the following sections we discuss some physical properties of the nanomaterials focusing briefly on the size-induced effects which serve as fundamental basis for understanding the applications and phenomena discussed in the next Chapters. For those interested in getting more deep into the subject, excellent textbooks are available [1].

2.2 Properties

2.2.1 *Structure and Morphology*

When going from a bulk material to a nanocrystal, the structure is affected thus leading to changes in lattice parameters and consequently the possibility of having new phases associated with the residual strain on the particles arises. This phenomenon has been observed for many materials either for spherical particles or thin films. X-ray diffraction technique is one of the most used techniques for evaluating structural properties of nanomaterials. In general, the diffraction pattern of crystalline nanoscale materials exhibit broadened and shifted peaks as compared to bulk and these changes are associated with both size and strain.

The effects of size and strain on the broadening of X-ray diffraction peaks in nanocrystals can be separated by constructing the so-called Williamson-Hall plots.

The full width half maximum β_{total} can be written as originating from the broadening due to size β_{size} plus the contribution from strain β_{strain} as follows [3]

$$\beta_{total} = \beta_{strain} + \beta_{size} = \frac{4\Delta d \sin\theta}{d \cos\theta} + \frac{k\lambda}{D\cos\theta}, \tag{2.1}$$

where Δd is the difference of d spacing corresponding to a given diffraction peak. In this model, the Scherrer constant k (in general assumed as 0.9 for spherical particles) is related to the shape of the nanocrystals, λ is the X-ray wavelength, and D is the average diffracted volume which is related to the nanocrystal size. The second term in equation 2.1 is the standard Scherrer´s model which is often solely used for evaluating the average crystal size. This is a good approximation only when the contribution of microstrain to the line broadening is negligible.

The first term in equation is related to microstrain which is defined as being the deviation of plane distance relative to their mean value, i.e. $\Delta d/d$. By constructing a plot of $\beta_{total}\ cos\ \theta$ vs $4\ sin\ \theta$ one obtains a line,whereby the intercept is related to inverse of crystal size and the slope yields the strain [1].

The role of reduced coordination of surface atoms on the nanocrystals is clearly unveiled in the structural properties. The main effect is lattice contraction by up to 3% as show in Figure 2.1 for some metallic nanocrystals [2]. The lattice contraction has been discussed in terms of a pressure which arises from the residual stress by using the Laplace´s law which reads [4]

$$\Delta P = \frac{4\gamma}{D} \tag{2.2}$$

where D is diameter of a liquid droplet and γ is the surface tension. For solid materials, the theory has been adapted and the strain can be expressed in terms of compressibility κ of bulk phase and surface tension γ. Following Wulff´s theorem [5] the deviation of lattice parameter can be written as

$$\frac{\Delta a}{a} \cong -\frac{4\kappa\gamma}{3D}. \tag{2.3}$$

By assuming that κ does not change as size is reduced, the parameter γ can be estimated using the $\Delta a/a$ *vs.* D^{-1} plot. Even a small change in size can lead to strain at GPa levels which is enough for stabilizing high pressure phase at ambient pressure. It was estimated an internal stress of 2.5 GPa for 40 nm $BaTiO_3$ nanocrystals [6]. In thin films, similar properties are observed but in this case is the substrate strain due to lattice mismatch which promotes the thin film to be highly stressed such as observed for the ferroelectric $PbTiO_3$ [7]. By comparing the Raman frequencies observed in the thin film with those observed for bulk at high pressure, it was possible to estimate that a 92 nm thick $PbTiO_3$ film has 1.7 GPa of strain. Other mechanisms are behind the lattice parameter changes in nanocrystals. For example, in case of CeO_2 it is observed as a lattice expansion when the nanocrystals size decreases and the phenomena have been identified as being due to the presence of oxygen vacancies [8].

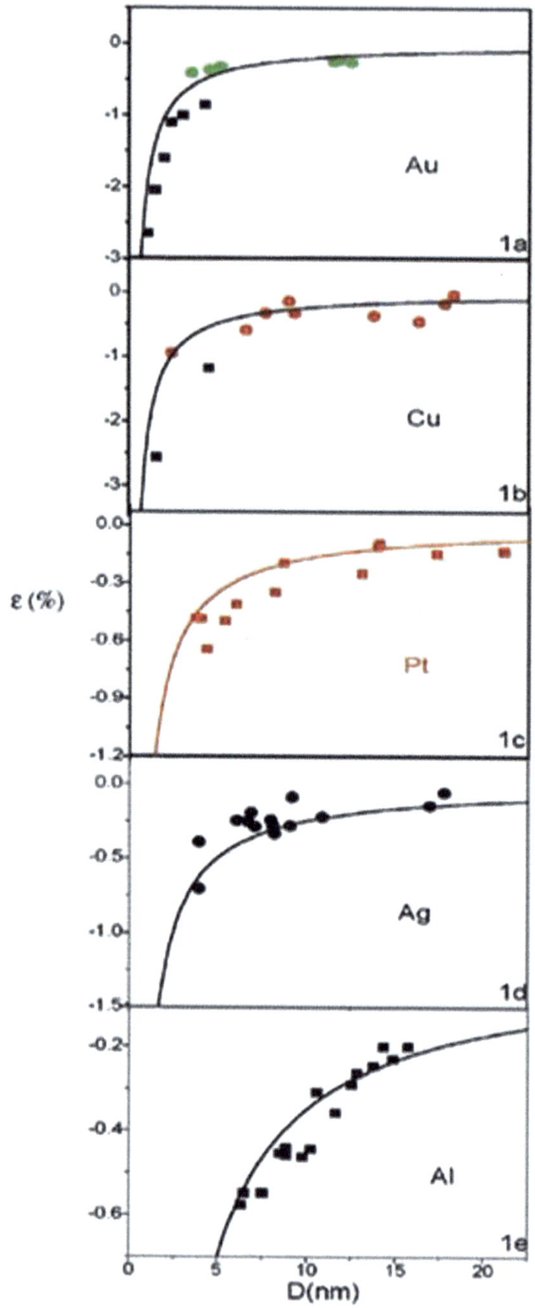

Fig. 2.1. Lattice contraction observed in metallic nanocrystals with fcc structure [2].

Morphology is another degree of freedom that can be used for tuning the nanomaterials properties. Up to date, researchers have been able to grow, using different techniques, a variety of morphologies with a very good control leading to obtain nanostructured materials with high degree of quality in both phase and morphology. Zinc oxide is a good example for illustrating the spectacular variety of morphologies with which a material can be prepared as seen in Figure 2.2. All the morphologies have the same wurtzite crystal structure. The different morphologies are formed as a result of the delicate balance between the energy from the polar charges, surface area, and elastic deformation. The mechanism responsible for such variety of forms is based on different growth rate along the three fastest growth directions and the polar surface phenomena which leads to a positive face due to Zn^{2+} and a negative face due to O^{2-} [9].

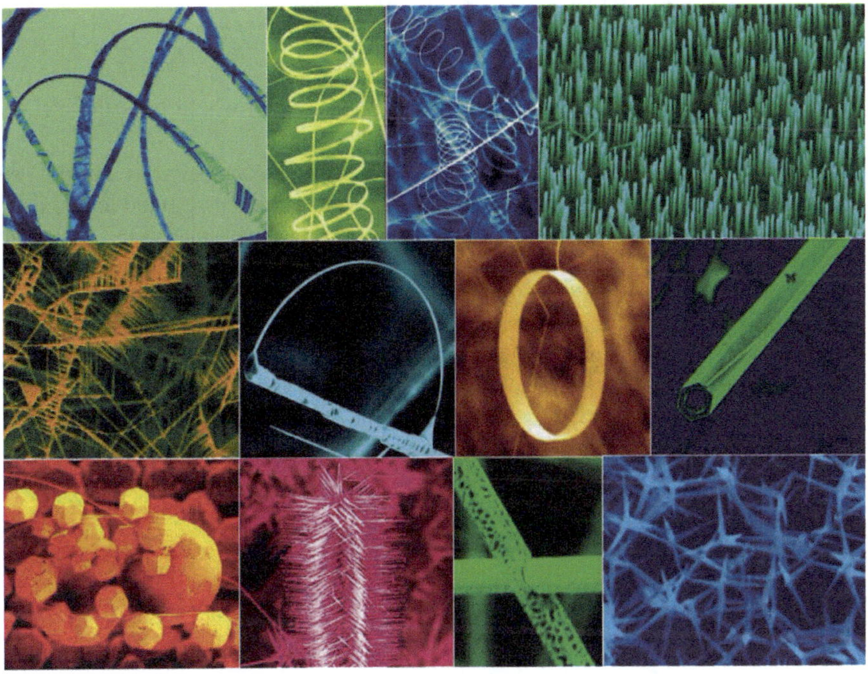

Fig. 2.2. Spectacular set of different morphologies observed in ZnO structures [9].

2.2.2 *Electronic and Optical Properties*

The electronic properties of a nanomaterial can be understood in terms of the electronic properties of the bulk. In the low-dimensional system, the electronic wavefunctions are constrained by quantum effects in the nanoscale directions. If the nanostructure has the same crystal structure as the bulk counterpart, the electronic states of the low-dimensional system is simply a subset of the electronic states of the bulk phase. Therefore, the wave vector components in the nanoscale directions can only take discrete values so that the wave vector components become quantized.

Electrons likewise behave quite differently in systems of different dimensionality. The electronic density of states g(E) (defined as the number of states with energy between E and E+dE) profiles for 3D, 2D, 1D and 0D systems are very different, as shown in Figure 2.3. The g(E) profile is crucial for many physical properties such as optical transitions and transport properties. In general the g(E) can be represented as

$$g(E)=\begin{cases} 0 & for\, E<E_j \\ C(E-E_j)^{\frac{d-2}{2}} & for\, E>E_j \end{cases} \tag{2.4}$$

where d denotes the dimension and it assumes the values 1, 2, and 3, respectively, for 1D, 2D and 3D systems. C is given by $\left(\frac{m}{2\pi^2\hbar^2}\right)^{1/2}$, $\left(\frac{m}{\hbar^2}\right)$ and $\left(\frac{m}{8\pi^4\hbar^2}\right)^{3/2}$ for 1D, 2D and 3D systems, respectively, where m is the electron mass and $\hbar$ is the Planck´s constant over 2π. E_j can be considered as a critical energy in the density of states [1]. For a 3D system E_j might correspond to an energy threshold for the onset of optical transitions, or a band gap energy E_g in a semiconductor. In 2D system such as quantum wells the E_i spectrum correspond to different band edge energy which arise from the confined states in one given dimension. For a 1D system, E_j would correspond to a van Hove singularity in the density of states occurring at each sub-band edge, where the magnitude of the electronic density of states becomes very large. In case of 0D systems the levels are completely discrete thus resembling a molecular system. In Figure 2.3 we show the g(E) profiles for different dimensionalities.

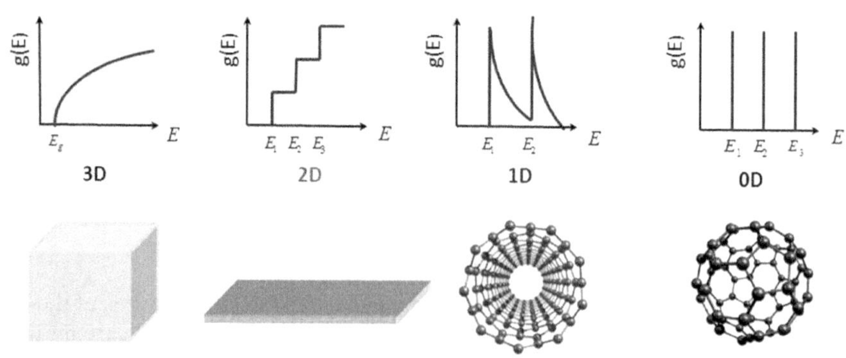

Fig. 2.3. Density of electronic state profiles for different dimensionality.

One of the most remarkable size-induced phenomena in nanoscale is the electronic behavior of the carbon nanotubes which we used here as a model system for illustrating size-induced effects.

Carbon nanotubes are one-dimensional materials, with cylindrical structures formed by tri-coordinated carbon atoms as schematically shown in Figure 2.4(a). These systems exhibit axial symmetry and spiral conformation or chirality [1] which is defined by the angle θ in Figure 2.4(b). Carbon nanotubes can be multi-wall (MWCNT) or single-wall (SWCNT) depending on the number of graphene sheets that are rolled up. The structure of the nanotubes is in first approximation understood in terms of the graphene strip as shown in Figure 2.4(b) where the points *O* and *A* are equivalent and define the circumference of the tube. These points are connected by a vector $\mathbf{C_h} = n\mathbf{a_1} + m\mathbf{a_2}$, where $\mathbf{a_1}$ and $\mathbf{a_2}$ are unit vectors for the graphene hexagonal lattice and *n* and *m* are arbitrary integer numbers that will fully define the nanotube structure. The translational symmetry is defined by the *B* and *B´* which are crystalographically equivalent to *O* and *A*, respectively, and connected by a translational symmetry vector.

Depending on the chiral vector ($\mathbf{C_h}$), a SWCNT can have three basic geometries based on the orientation of hexagons relative to nanotubes axis: (i) armchair with $\theta = 30^{\circ}$ (n = m); (ii) zigzag with $\theta=0^{\circ}$ ($n \neq 0$, m = 0) and (iii) chiral with $0^{\circ} < \theta < 30^{\circ}$ ($n \neq m$ and $m \neq 0$) [10]. In Figure 2.4(c) we illustrate the three kinds of nanotubes according to their symmetry, the so called *chiral* (11,5) and *achiral* (16,0) and (8,8) which are named zigzag and armchair, respectively.

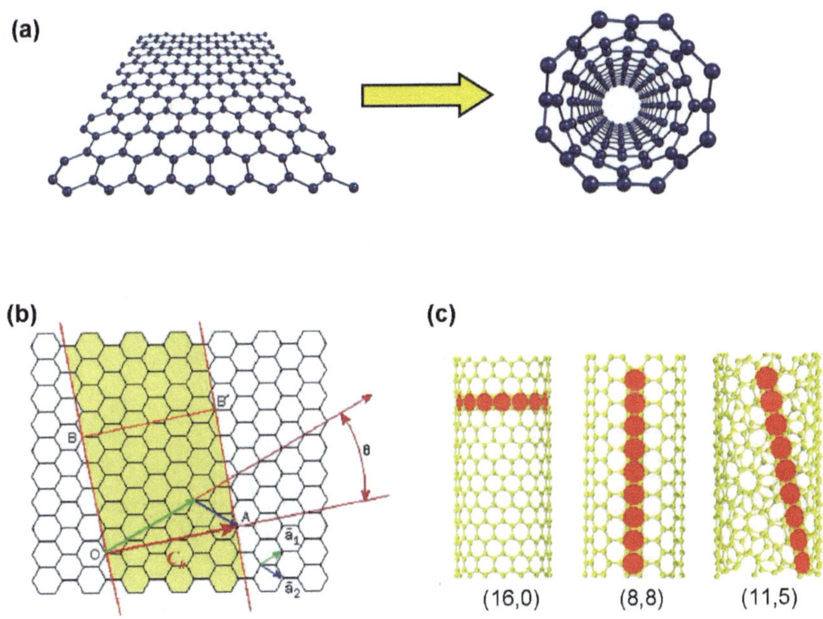

Fig. 2.4. (a) Schematic diagram showing how a carbon nanotube is conceptually built starting from a graphene strip. (b) The chiral vector $\mathbf{C_h}=n\mathbf{a_1}+m\mathbf{a_2}$ definition in terms of basis vectors for the hexagonal lattice. In this case n=4 e m=2. (c) Carbon nanotubes with $\theta=0^0$ (16,0) $\theta=30^0$ (8,8) and $\theta=19.0^0$ (11,5) [10].

Carbon nanotubes are not a planar hexagonal carbon structure and the resulting sp^2 hybridization of the carbon atoms is not perfect, since the directions of orbitals suffer distortions due to the curvature effects of the surface of the tube. The resulting hybridization is $sp^{2+\delta}$ (with $0 < \delta < 1$), leading to new electronic properties. The boundary conditions around the circumference lead to quantization of the electronic states in this direction while the wavector value along the nanotube axis is in continuous states as expected for 1D systems.

Carbon nanotubes have a unique relationship between their geometries and their electronic properties. Due to the symmetry of graphene sheet and by imposing the quantization of the wave-vector along the circumferential direction, a very simple rule based only on structure is obtained for classifying nanotubes regarding their conductive behavior. If $n - m$ is multiple of 3 the nanotube is metallic, otherwise it is semiconducting [10].

The density of electronic state for a semiconducting and a metallic nanotube is shown in Figure 2.5. The electronic transitions between van Hove singularities in the valence and conduction bands are very strong and named as E_{ii} $(i=1,2,3\ ...)$ and optical and transport properties strongly depend on these resonant states. In the case of semiconducting nanotubes the first transition E_{11} is identified as the *gap* and it roughly depends on the tube diameter d_t as $E_{11} \sim d_t^{-1}$. The departure from this scale law is due to chirality effects [11]. This simple picture discussed above does not take into account the curvature effects which change this picture especially for nanotubes with diameters below 1 nm.

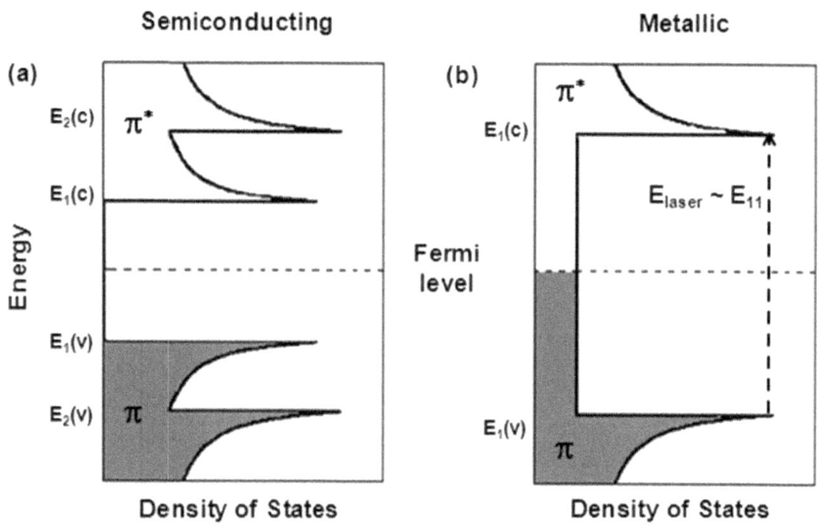

Fig. 2.5. Density of electronic states for a semiconducting (a) and a metallic nanotube (b). The Fermi level is illustrated by the dashed horizontal line which separates the occupied π and unoccupied π^* states.

The optical properties of carbon nanotubes are special and when they are individualized is possible to measure well-resolved excitation-emission optical maps and associate each peak to a given (n,m) semiconducting carbon nanotubes. The nanotubes are individualized using wrapping molecules such as surfactants [12]. The absorption and emission are very strong due to the van Hove singularities in the density of electronic states. The accepted mechanism involved in the observed optical phenomena in nanotubes is described by considering the excitonic effects [13]. The emission wavelength depends on the effective dielectric constant of environment and this property has been used for making ultra-sensitive optical sensors [14]. The photoluminescence is so strong that it is possible to observe and control its intensity stepwise as the nanotubes undergo a chemical reaction [15]. Carbon nanoribbons are also considered 1D and their gap scales approximately as the inverse of ribbon width, $E_{gap}{\sim}W^{1}$.

The band gap in semiconducting nanocrystals such as CdS is strongly dependent on size which is a good example for illustrating the size-induced properties. The band gap in quantum dots is approximately described by a D^{-2} dependence in analogy with the standard problem of a particle in a box.

Metallic nanoparticles are very important in nanotechnology as they exhibits interesting physical properties which have found applications in different fields varying from optics to catalysis. The nearly free conduction electrons in these particles undergo collective excitation called *plasmon resonances* which play a major role on the optics of these systems. The plasmon energy is strongly dependent on the size and shape of the nanoparticles and has been used as the fingerprint of metallic nanoparticles [1].

2.2.3 Vibrational Properties

Vibrational properties are considerably affected by the size reduction and these size-induced phenomena have been exploited for accessing nanomaterials properties such as particle size, morphology, disorder, etc. In particular, Raman spectroscopy has been one of the most used techniques for measuring vibrational properties of different nanomaterials. Raman spectroscopy is based on the inelastic light scattering by the lattice vibrations (phonons) and in bulk materials the scattering obeys the momentum selection rule which states that only $q \sim 0$ (q is the phonon wavevector) is allowed.

In the case of a nanocrystal, the $q \sim 0$ wavevector selection rule is breakdown because of uncertainty principle and broad and asymmetric modes are observed because phonons belonging to the interior of the Brillouin zone $q \neq 0$ also contribute to the observed Raman profile. The peak profile of the nanocrystal depends strongly on size and on the phonon dispersion relation of the bulk material. The phonon confinement model predicts that the Raman profile can be described by the following equation [16]:

$$I(\omega) = \int \frac{|C(q)|^2 \, d^3q}{[\omega - \omega(q)]^2 + (\Gamma_o/2)^2} \tag{2.5}$$

where Γ_o is the line width of the Raman peak and $\omega(q)$ is the phonon dispersion curve for the bulk counterpart. The C(q) coefficient depends on the nanocrystal morphology and carries information on the relevant size parameters as listed in Table 2.1.

Table 2.1. Fourier coefficients for different morphologies. [16]

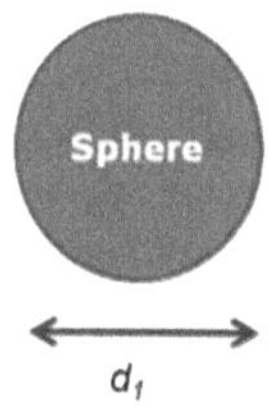	$\lvert C(q)\rvert^2 \approx \exp\left(-\frac{q^2 d_1^2}{16\pi^2}\right)$
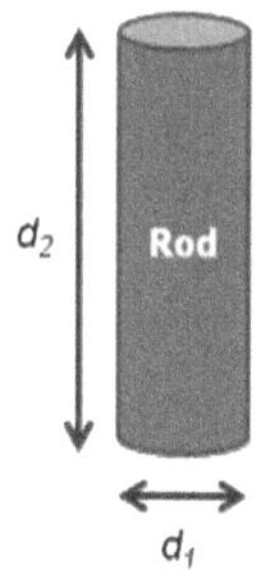	$\lvert C(q)\rvert^2 \approx \exp\left(-\frac{q_1^2 d_1^2}{16\pi^2}\right)\exp\left(-\frac{q_2^2 d_2^2}{16\pi^2}\right)\left\lvert 1 - erf\left(-\frac{i q_2 d_2^2}{\sqrt{32\pi}}\right)\right\rvert^2$
	$\lvert C(q)\rvert^2 \approx \exp\left(-\frac{q_1^2 d_1^2}{16\pi^2}\right)\left\lvert 1 - erf\left(-\frac{i q_1 d_1^2}{\sqrt{32\pi}}\right)\right\rvert^2$

The experimental Raman profile is fit to the above equations and the relevant dimensions are determined using Raman spectroscopy. The values found using this technique for several systems are in good agreement with those obtained by direct methods such as transmission electron microscopy.

In Figure 2.6 we show how the frequency of the vibrational mode in CeO_2 nanocrystal can be correlated with the particle average size [17]. This technique is interesting because it probes a very large number of nanoparticles (the laser spot is about 1μm) thus allowing one to measure a representative distribution of the nanocrystal size present in the sample.

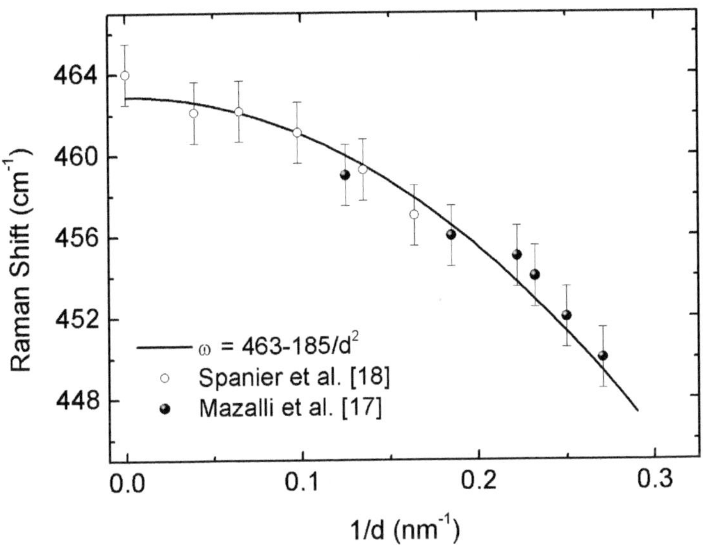

Fig. 2.6. Diameter dependence of Raman mode for CeO_2 nanocrystal [17].

2.2.4 Magnetic Properties

Magnetic properties are very important for technology and can be exploited in many aspects ranging from recording to drug delivery systems. The size induced phenomena properties is one of the most striking phenomena of the mater at the nanoscale and clearly show how the surface atom coordination play the major role.

The magnetic properties come from a genuine quantum effect related to spin interaction. The exchange interaction force is so strong below the critical Curie temperature, T_C, that the material experiences a magnetization alignment even when the external magnetic field is zero. Below T_C the magnetic moments experiences a long range order and they tend to align in a given direction resulting in a net magnetization. In practice, the net magnetization is null for almost all materials because they are composed of magnetized regions called domains and these domains have a random distribution of magnetization direction.

The materials can be grouped into ferromagnetic, ferrimagnetic, anti-ferromagnetic and paramagnetic. Each group has a different behavior when a magnetic field is applied. A ferromagnetic material is characterized by the alignment of the magnetic moments leading to a spontaneous magnetization even in the absence of an external magnetic field. The anti-ferromagnetic and ferrimagnetic states arise when the magnetic moments arrangement consist of two lattices with magnetic alignment in opposite directions. In the anti-ferromagnetism these lattices has exactly the same magnetization values and the total magnetization is zero while for ferrimagnetic the lattices have different magnetization values thus resulting in a net magnetization. The paramagnetic states are characterized by the alignment of the magnetic moments when an external field is applied.

The main physical quantity that is usually measured in magnetism is the magnetic susceptibility which couple the external magnetic field H with the magnetization M, i.e., $M=\chi H$. The χ^{-1} scales with T being written as

$$\chi^{-1}(T) = \frac{T-\theta}{C}, \tag{2.6}$$

where C is the Curie constant which is related to the effective atomic magnetic moment. Above the critical temperature, the sign of θ defines the magnetic order alignment for $T > T_C$. The Curie paramagnetism is defined for $\theta = 0$ and Curie-Weiss paramagnetism for $\theta > 0$. The anti-ferromagnetic alignment is obtained for $\theta < 0$ where the critical temperature is named Neel temperature T_N. These properties are observed for bulk materials.

Small magnetic particles exhibit unique phenomena such as quantum tunnelling of magnetization and superparamagnetism. The former phenomenon is observed

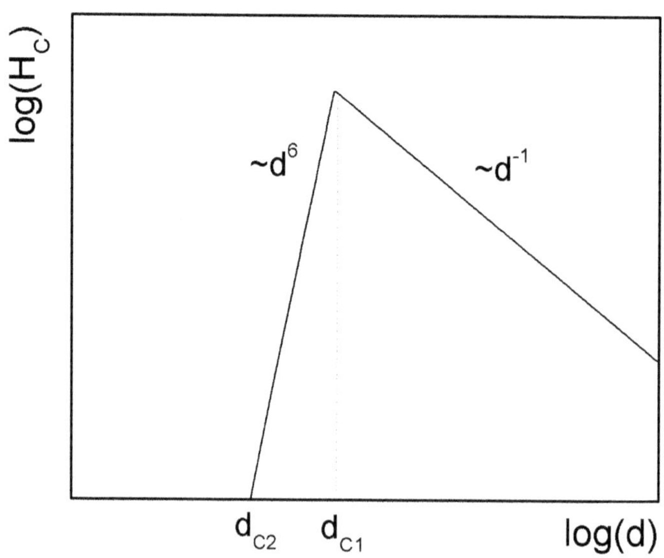

Fig. 2.7. Diameter dependence of coercive field as proposed by Herzer [19].

in molecular magnets and it is not discussed in this chapter. Superparamagnetism is the main phenomena observed at the nanoscale. When going from a bulk ferromagnetic material to a nanoparticle, the typical hysteresis curve disappears and the system enters in a regime whereby large reversible magnetization is observed. It should be pointed out that the superparamagnetism has been observed below 4K for metal nanoparticles (average size of 2.5 nm) that usually are not magnetic in bulk form such as gold and palladium. The temperature at which the hysteresis loop disappears is called the blocking temperature. By decreasing the size, the energy needed for rotating the magnetization out of the easy direction is reduced (the so called anisotropy energy) and when it becomes lower than k_BT the superparamagnetism phenomena arises.

The size dependence of coercive field H_C was described by Herzer [19] which is schematically shown in Figure 2.7. The critical diameter d_{c1} is the critical exchange length L_{ex} which defines the limit for multi-domain ($d > d_{c1}$) and single domain ($d < d_{c1}$) configuration. The region below critical diameter d_{c2} defines the superparamagnetic regime. Experimental results on nanocrystalline alloys confirm the diameter dependence shown in Figure 2.7 [19].

Magnetic nanocomposites are important systems for applications. These systems consist of magnetic particles dispersed into non-magnetic matrices such as a porous vycor glass (PVG). It has been shown that Fe_2O_3 nanoparticles dispersed into PVG exhibit a superparamagnetic behavior. The zero field cooling (ZFC) curve of a non-interacting system constituted by single domain nanoparticles can be calculated considering the distribution of particle volumes ($f(V)$) as

$$\frac{M_{ZFC}(H,T)}{M_S} = \frac{M_S^C H}{3K}\frac{1}{\Omega}\int_{V_b(T)}^{\infty} f(V)dV + \frac{M_S^C H}{3k_B T}\frac{1}{\Omega}\int_0^{V_b(T)} Vf(V)dV \qquad (2.7)$$

where $\Omega = \int_0^{\infty} Vf(V)dV$, $V_b(T) = 25k_BT/K$, H is the applied field, M_S^C is the saturation magnetization of the nanoparticle, k_B is the Boltzmann constant, K is a parameter that accounts for the effective anisotropy of the nanoparticles, M_S is the sample saturation magnetization, and T is the absolute temperature [20,21]. The first term in Equation 2.7 comes from the blocked particle contributions while the second comes from the superparamagnetic ones. The field cooling (FC) magnetization curve can be calculated by using an expression similar to Equation 2.7 [20]. The ZFC and FC curves of the samples can be studied by considering a log-normal distribution of volumes, M_S^C = 420 emu/cm^3 (bulk value), and using the value of $K = 5.9\text{x}10^5$ erg/cm^3 [22]. Figure 2.8 shows the calculated magnetization curves for ZFC and FC for different samples as solid and dotted lines, respectively.

As can be seen in Figure 2.8, The ZFC curves are well described by the superparamagnetic behavior, even for the sample with a large number of IDC, thus suggesting that particle agglomeration is not significant in such samples. The average particle size $\overline{D}$ increases with the increasing number of IDC (impregnation decomposition cycles used in the sample preparation [23]) in agreement with the analysis by Transmission Electron Microscopy (TEM) [23]. For γ-Fe_2O_3/PVG with 8 IDC, the $\overline{D}$ determined by TEM and magnetic analysis show great agreement and was found to be around 5.2 nm.

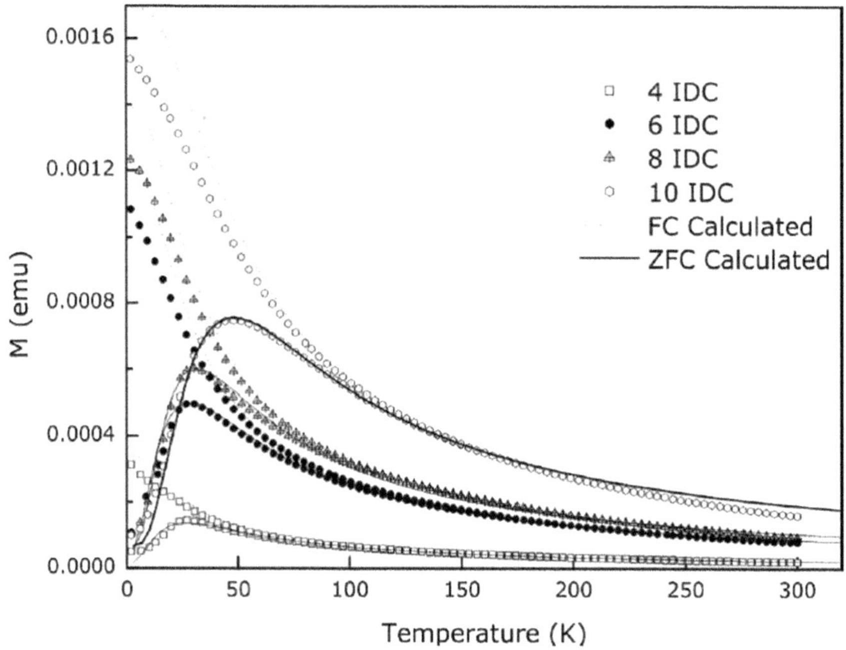

Fig. 2.8. Temperature dependence of the ZFC solid lines and FC dotted lines magnetization curves for γ-Fe_2O_3 /PVG with different numbers of IDC [23].

2.2.5 *Ferroelectric Properties*

Ferroelectricity phenomenon is very important in fundamental science and applications as well. It is a cooperative phenomenon associated with a spontaneous electric polarization moment which arises from structural distortions in the crystal lattice. Similar to ferromagnetism, the spontaneous alignment of electric dipoles happens below a critical temperature T_c and it happens along with a structural phase transition. This critical temperature depends on particle size as shown for $PbTiO_3$ particles by Ishikawa et al [24] (see Figure 2.9). The T_C experimental data was adjusted with the following equation

$$T_C(D) = \frac{T_C^{bulk} - \Phi}{D - D_0} \tag{2.8}$$

where D is the particle size, Φ is a fitting parameter, D_0 is the particle size for T_C=0 K, and T_C^{bulk} is the ferroelectric-paraelectric transition temperature in bulk [25]. Following this model, below the critical diameter where the polarization vanishes is called superparaelectricity regime in analogy to superparamagnetism. The critical size for $PbTiO_3$, $Bi_4Ti_3O_{12}$, and $PbZr_{0.3}Ti_{0.7}O_3$ is 9.1nm, 44 nm, and 13 nm, respectively [24,27,28].

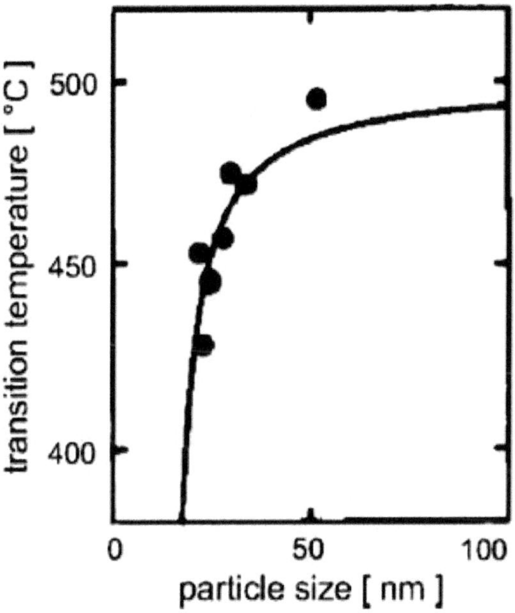

Fig. 2.9. Particle size dependence of ferroelectric-paraelectric phase transition temperature in $PbTiO_3$ nanoparticles [24].

The physical basis for polarization ceasing is discussed in terms of free energy calculations. It has been shown that in introducing a polarization gradient for polarization and a term related to surface in the free energy, the cubic (paraelectric phase) symmetry is preferred in detriment of tetragonal (ferroelectric phase) symmetry. Another way of viewing this has to do with the role of surface atoms. The surface strain decreases T_C values at the surface layers and as the size decreases they dominate and the whole particle become cubic. Addressing the superparaelectricity phenomenon is key because of its applications.

2.3 Concluding Remarks

Some basic properties of nanomaterials were discussed and correlated with the role played by the surface atoms and the quantum-induced effects (for both electrons and phonons) in nanoscale when compared with their bulk counterparts. One consequence of these effects are the structural and morphological properties that change considerably when bulk structures goes to nanoparticles leading to changes in lattice parameters and consequently the possibility of having new phases and morphologies. Quantum confinement is responsible for unique electronic properties of one-dimensional systems such as carbon nanotubes which can show metallic or semiconductor behavior depending on the geometry. Fundamental phenomena such as ferromagnetism and ferroelectricity are also strongly dependent on size and the nanomaterials are allowing one to access fundamental physics of these phenomena.

Acknowledgments. The authors acknowledge funding from Brazilian agencies CNPq, CAPES grant PROCAD 068/2007, FUNCAP, FAPERGS, INCT NANOBIOSIMES and INCT Nanomateriais de Carbono.

References

[1] (a) Roduner E (2006) Nanoscopic materials: size-induced phenomena. RSC Publishing, Cambridge; (b) Cao G (2004) Nanostructures & Nanomaterials: Synthesis, properties and applications. Imperial College Press, London
[2] Jiang Q, Liang LH, Zhao DS (2001) Lattice contraction and surface stress of fcc nanocrystals. J Phys Chem 105:6275-6277
[3] Williamson GK, Hall WH (1953) X-ray line broadening from filed aluminium and wolfram. Acta Metall 1:22-31
[4] Solliard C, Flueli M (1985) Surface stress and size effect on the lattice-parameter in small particles of gold and platinum. Surf Sci 156:487-494
[5] Herring, C (1953) Structure and properties of solid surfaces. In: Gomer R and Smith CS (eds) University of Chicago, Chicago
[6] Uchino K, Sadanaga E, Hirose T (1989) Dependence of the crystal-structure on particle size in barium titanate. J Am Ceram Soc 72:1555-1558
[7] Valim D, Souza Filho AG, Freire PTC, Mendes Filho J, Guarany CA, Reis RN and Araujo EB (2004) Evaluating the residual stress in $PbTiO_3$ films obtained from a polymeric method. J Phys D: Appl Phys 37:744–747
[8] Zhou XD and Huebner W (2001) Size-induced lattice relaxation in CeO_2 nanoparticles. Appl Phys Lett 79:3512-3514
[9] Wang ZL (2004) Nanostructures of zinc oxide. Materials Today 7:26-33
[10] Saito R, Dresselhaus G and Dresselhaus MS (1998) Physical Properties of Carbon Nanotubes. Covent Garden Imperial College Press, London
[11] Saito R, Fujita M. Dresselhaus G and Dresselhaus MS (1992) Electronic Structure of Chiral Graphene Tubules. Applied Physics Letters 60:2204-2206
[12] O'Connell MJ, Bachilo SM, Huffman XB, Moore VC, Strano MS et al (2002) Band gap fluorescence from individual single walled carbon nanotubes. Science 297: 593–96
[13] Wang F, Dukovic G, Brus LE, Heinz TF (2005) The optical resonances in carbon nanotubes arise from excitons. Science 308:838–41
[14] Heller DA, Jeng ES, Yeung TK, Martinez BM, Moll AE, Gastala JB, Strano MS (2006) Science 311:511
[15] Siitonen AJ, Tsyboulski DA, Bachilo SM, Weisman RB (2010) Surfactant-Dependent Exciton Mobility in Single-Walled Carbon Nanotubes Studied by Single-Molecule Reactions. Nano Lett 10:1595–1599

[16] Richter H, Zhang ZP, Ley L (1981) The one phonon Raman spectrum in microcrystalline silicon. Solid State Commun 39:625-629
[17] Mazali IO, Viana BC, Alves OL, Mendes Filho J, Souza Filho AG (2007) Structural and vibrational properties of CeO_2 nanocrystals. J Phys Chem Sol 68:622–627
[18] Spanier JE, Robinson RD, Zhang F, Chan SW, Herman IP (2001) Phys. Rev. B 64. 245407
[19] Herzer G (1990) Grain-size dependence of coercivity and permeability in nanocrystaline ferromagents. IEEE Trans Magn 26:1937-1402
[20] Respaud M, Broto JM, Rakoto H, Fert AR, Thomas L, Barbara B, Verelst M, Snoeck E, Lecante P, Mosset A, Osuna J, Ely TO, Amiens C, and Chaudret B (1998) Surface effects on the magnetic properties of ultrafine cobalt particles. Phys Rev B 57: 2925-2935
[21] Hansen MF and Morup S (1999) Estimation of blocking temperatures from ZFC/FC curves. J Magn Magn Mater 203:214-216
[22] Nunes WC, Cebollada F, Knobel M, and D. Zanchet (2006) Effects of dipolar interactions on the magnetic properties of gamma-Fe_2O_3 nanoparticles in the blocked state. J Appl Phys 99:08N705
[23] Cangussu D, Nunes WC, Correa HLS, Knobel M, Macedo WAA, Alves OL, Souza Filho AG, Mazali IO (2009). g-Fe_2O_3 nanoparticles dispersed in porous Vycor glass: a magnetically diluted integrated system. Journal of Applied Physics 105:013901-1-013901-7
[24] Ishikawa K, Yoshiwawa K, Okada N (1988) Size effect on the ferroelectric phase-transition in $PbTiO_3$ ultrafine particles. Phys Rev B 37:5852-5855
[25] Zhong WL, Jiang B, Zhang PL, Ma JM, Cheng HM, Yang ZH, Li LX (1993) Phase transition in $PbTiO_3$ ultrafine particles of different sizes. J Phys: Condens Matter 5:2619
[26] Du YL, Zhang MS, Chen Q, Yuan ZR, Yin Z, Zhang QA (2002) Size effect and evidence of a size-driven phase transition in $Bi_4Ti_3O_{12}$ nanocrystals. Solid State Commun 124:113-118
[27] Meng JF, Katiyar RS, Zou GT, Wang XH (1997) Raman Phonon Modes and Ferroelectric Phase Transitions. Phys Stat Sol (a) 164:851-862

Abbreviations

0D – zero-dimensional

1D – one-dimensional

2D - two-dimensional

3D - bulk

FC - field cooling

IDC – impregnation decomposition cycle

MWCNT – multi-wall carbon nanotubes

SWCNT - single-wall carbon nanotubes

ZFC - zero field cooling

T_C - Curie temperature

TEM – Transmission Electron Microscopy

PVG – porous vycor glass

3 Nanomagnetic Materials

Silvana Da Dalt[2], Priscila Chaves Panta[2], and Juliano Cantarelli Toniolo[1]

[1] Instituto Federal de Educação, Ciência e Tecnologia do Rio Grande do Sul, 95012580, Caxias do Sul, Brazil
[2] Departamento de Engenharia de Materiais, Universidade Federal do Rio Grande do Sul, 90035190, Porto Alegre, Brazil
E-mail: silvana.da.dalt@ufrgs.br

Abstract. The dynamic development of technological civilization depends increasingly on the development of materials engineering, which seeks non-conventional materials when it comes to their physical, chemical and mechanical properties. The search for new magnetic materials, e.g., brought with it the development of metallic-based materials, such as iron, combined with unique nanocrystalline structures and properties that were never seen or had not yet been obtained. This fact is reflected in research in this area, whose focus has changed from the microcrystalline to the nanocrystalline size. Thus, its applications in the fields of biomedicine, molecular biology, biochemistry, diagnosis, catalysis, and various other industrial applications are increasingly being researched. These materials can be used both as magnetic particles and as magnetic fluids. There are several industrial applications of magnetic nanoparticles, such as magnetic seals in motors, magnetic inks for bank checks, magnetic recording media, among others.

Keywords: nanomagnetic particles, biomedicine applications, iron oxide.

3.1 Introduction

The need to obtain new technologies has stimulated the development of new materials, where properties can be controlled and even be substantially better than previously projected. In this regard, one of the main classes of materials is magnetic materials with the strategic interest. At this juncture, the nanomagnetism is a specialty that deals with magnetic phenomena especially present in structures with dimensions below the micrometer, i.e. nanometer (10^{-9} m). This is a field of nanoscience with a rich and intriguing physics behind and at the same time, it is intimately aligned with new technological challenges, which began with the discovery of giant magnetoresistance effect [1].

Currently, nanotechnology, in which scientists build tiny particles combining inorganic and organic materials, is taking the boundary of the scientific field that progresses at a frightening speed. Important advances in diagnosis, therapy, molecular biology and bioengineering are underway in biomedical nanotechnology. If one hand is far the feasibility of building large part of nanodevices imagined, today it is quite plausible the use of nano-drugs, such as those based on liposomes and nanoparticles, for the placement and treatment of many diseases. These systems are especially interesting if they are constructed from magnetic nanoparticles [2].

There are several industrial applications for magnetic nanoparticles such as magnetic seals in engines, magnetic inks for bank checks, magnetic recording media for data storage, among others. With increased understanding and control of nanomagnetic phenomena, several important applications have been proposed recently, such as the production of ultra-strong permanent magnets, information storage systems with high magnetic density, nano-biomagnetic sensors, electronics based on electron spin, etc. [3].

The nanoscale size effects of confinement and/or surface are replaced by a great influence on the properties of the system, which can have both positive and negative consequences depending on the type of application [4]. One example is the superparamagnetic regime that arises when the reduction in size of the nanoparticles is sufficient for the magnetic moment that does not become stable in the preferred magnetic axis during a typical measurement time [5,6]. For applications in recording media, this issue has been carefully addressed, because it represents a limiting factor in even the smallest "bit" of information possible. Moreover, the influence of magnetic interactions in these systems can also be another limiting factor.

The search for new magnetic materials, such as iron, e.g., led the development of metallic materials with a nanocrystalline structure and non-conventional properties. The process improvement in the control of production and characterization of these systems make possible an excursion to a set of physical information of a hitherto unknown world. This has entailed to numerous benefits, both in the technological field and in the development of understanding of phenomena involved in this scale, which increases with the control degree of the fine structure of matter.

This chapter aims to provide information about recent developments and available nanomagnetic systems commonly studied (nanoparticles, thin films and magnetic fluids) in the literature. Possible technological applications will be addressed in the context of five particular groups:

1. Biomedicine
2. Magnetic Storage Media
3. Particle Separation by Magnetic Action
4. Nanosensors and Devices
5. Catalysts and Pigments

3.2 Magnetic Nanoparticles Applications in Biomedicine

Excellent studies have been performed in numerous ways, such as in the biomedical applications of magnetic materials. In general, these applications involve the administration inside the body (in vivo) and outside the body (in vitro). In vivo applications can further be classified as therapeutic (drug delivery and hyperthermia) and diagnostic (nuclear magnetic resonance imaging), while applications in vitro comprise essentially of diagnostics (separation / selection) [7]. Consequently, there has been an increase in the number of scientific research aiming to develop, characterize and improve nanomagnetic materials with drug delivery functions inside the human body, both in relation to surface functionalization of molecules with high specificity and in relation to the best properties of the magnetic particles.

These materials play an important role in modern technology since they find applications in a wide range of industrial products and processes in various sectors. These applications include permanent magnets which are used in door locks, electric motors, electronic scales, sensors, etc. Iron oxides, or materials containing it, are the basis for most of these applications. Consequently, iron oxides have received increasing attention because of their extensive applications in magnetic recording media, catalysts, pigments, sensors, electromagnetic and optical devices, among others.

There is, therefore, a wide variety of properties that distinguish one magnetic material from another for a given application. Among these properties, we can observe a variety of structures, generating polymorphic materials, in addition to, other physical characteristics, like the magnetic character and chemical properties, such as the bonds between atoms or ions, changes in particle size and the degree of hydration. By definition, particles of superparamagnetic iron oxide are usually classified according to a size below 10 nm of the particles, since the order of this size is formed by mono-domain magnetic particle, causing an obstruction, then the saturation magnetization, or that is, the complete pairing of the spins [8]. It is also said that iron oxides as magnetic materials for biomedical applications are the subject of frequent investigations due to their superior biocompatibility compared with other magnetic materials, both based on oxides or pure metals. There are several types of iron oxides found in nature and which can be synthesized in the laboratory, but presently maghemite (Fe_2O_3) and magnetite (Fe_3O_4) are able to meet the requirements for biomedical applications.

These requirements include the attainment of high values of magnetic moments, chemical stability under physiological conditions, low toxicity, not to mention the possibility of low production cost of the material and the many available synthesis procedures for the preparation of these materials. The degree of crystallinity, and dispersability in terms of the size and shape of nanoparticles are critical parameters that affect their performance in the therapeutic and diagnostic techniques such as MRI and hyperthermia. These parameters are strongly correlated with the approach taken for their preparation. There are studies aiming to apply magnetic nanoparticles coated with powdered carbon structures. These nanoparticles can be coated with carbon nanotubes, forming a sort of wrapper, because besides preserving the magnetic characteristics of nanoparticles, the carbon can also bind to various substances such as metals, catalysts, polymers and organic compounds. This feature allows, e.g., applications in medicine, with the material carrying the active ingredients inside the body [8].

3.2.1 Drug Delivery

New discoveries about nanoparticles with magnetic properties are continually developing a promising field for applications in medicine and the pharmaceutical industry. Magnetic nanoparticles are being actively investigated in order to establish the drug delivery generation. The systems for controlled release of drugs involve different, multidisciplinary aspects and can contribute greatly to the advancement of human health. These systems offer many advantages when compared to other systems of conventional dosing [9].

Consequently, there has been an increase in the number of laboratories seeking to develop, characterize and improve the nanomagnetic materials that are going to be used as porters (drug molecules or simple magnetic moment) inside the human body, both in regard to surface functionalization with molecules of high specifity, and in relation to the best properties of the magnetic particles. Many studies are also occurring in the field of nanoparticles to use these systems for drug delivery. These are designed based on new strategies for the placement of active ingredients, which include important applications of polymer science and of surfactants solutions and the preparation of colloidal species, administration of DNA vaccines, besides the use of transdermal techniques [9]. The potential for developing systems for distributing drugs based on the use of nanoparticles is based on significant advantages such as:

- The ability to target specific sites in the body without significant immobilization of the bioactive species;
- Reduction in the amount of the drug (toxicity), since the target is established and only this region will be affected with longer time spent in circulation;
- Secure administration with a reduction of drug concentration (fewer doses) in places that are not the primary target, but close to it, thus minimizing side effects and inflammatory reactions;
- Enhanced therapeutic efficacy, with gradual and controlled release of drug from the matrix degradation;
- Both hydrophilic and lipophilic substances can be incorporated;
- Varied nature and composition of the vehicles and, contrary to what one might expect, there is no predominance of mechanisms of instability and decomposition of the drug (premature bio-inactivation).

The greatest disadvantage of most conventional chemotherapy for the treatment of cancer, whose administration is intravenous, is its general systemic distribution throughout the body. This technique results in several known side effects of chemotherapy. These chemical drugs attack not only tumor cells but also healthy cells. To overcome this disadvantage, some types of magnetic nanoparticles can be employed for this purpose. The particles loaded with drugs are sent to the destination with the aid of an external source, like a magnet. These drugs are then released into the desired area [10]. Figure 3.1 shows a scheme of the drug delivery technique.

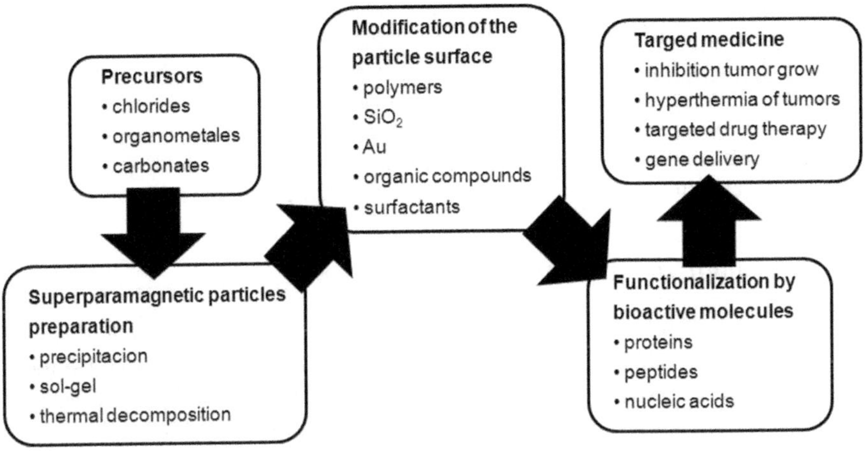

Fig. 3.1. Scheme of the magnetic particles preparation used in the drug delivery technique. Adapted from [10].

The use of such magnetic nanoparticles, however, depends largely on the processes of synthesis, so that then the best structural and magnetic characteristics and conditions can be selected, such as particle size, homogeneity of the material obtained, surface area and optimization of magnetic properties. The capacity to increase the magnetization of these particles is beneficial for enabling the manipulation of drug distribution. The nanoparticles should also be small enough to be superparamagnetic, thus avoiding clumps after the cessation of external magnetic field and that they remain in circulation without being removed by the body's natural filters such as the liver or the immune system [11]. Crystalline, polymorphous iron oxides (Fe_2O_3 and Fe_3O_4), such as hematite, maghemite, magnetite and others are most frequently used as sources of magnetic materials [12]. However, maghemite and magnetite are of greater interest to bioapplications [13]. Some pure metals such as Fe and Co have been chosen as magnetic materials, since it is not uncommon for them to present several advantages compared to iron oxides, such as high saturation magnetization. However, Fe and Co prove to be more susceptible to oxidation in aqueous solutions and also present lack of compatibility when compared to iron oxides.

The use of magnetite nanoparticles has been studied for applications in the development of immunoassays, contrast agents for magnetic resonance imaging, target drug delivery vehicles (drug delivery), as well as magnetic hyperthermia, where the superparamagnetic behavior of these nanoparticles, used for treatment in fighting cancer cells, when exposed to an alternating magnetic field can be used to heat the tumor cells to 41 °C and 45 °C, where the damage of normal tissue is reversible, whereas the tumor cells are irreversibly damaged [14].

In these applications, the nanoparticles should exhibit superparamagnetic behaviour, because if there was any remnant magnetization, there could be an agglomeration of particles, which should be avoided inside the body to prevent blockage of blood vessels. Another alternative is the encapsulation of magnetic nanoparticles with polymers, also used as drug delivery, in order to optimize the stability of the nanoparticles against oxidation, as demonstrated in several examples [15,16]. Reshmi et al. reported the method of preparation of such materials using iron carbonyl and biodegradable polymers (cellulose acetate phthalate hydrogen) [17]. The polymer helps in transporting and distributing the drug during its degradation [18].

3.2.2 Magnetic Hyperthermia

The use of hyperthermia in the treatment of malignant tumors is as old as medicine itself. For example, Hippocrates, the father of medicine, proposed that the superficial tumors (such as skin tumors) should be cauterized by the application of a hot iron. Thus, it follows that magnetic hyperthermia is a therapeutic procedure applied to provide an increase in temperature in a given region of the body that is affected by a neoplasm. Once detected the presence of tumor cells through their aggregation with magnetic particles, which act like tiny magnets, like red blood cells in an attempt to trick the immune system, and causing them to vibrate by the action of an external magnetic field to the body. Through this vibration the magnetic particles dissipate heat onto the tumor cells, causing lysis and death. Since that there have been numerous publications describing a variety of systems using different types of magnetic materials, different field strengths and frequencies and different packaging and distribution methods of the particles [23].

The process of magnetothermocytolysis, which is defined by cell death due to absorption of magnetically generated heat, is therefore an application within the nano-biotechnological processes, because it leads to the destruction of specific cancer cells without affecting normal cells from surrounding tissues, which is currently the biggest concern in radiation therapy, even when the most sophisticated linear accelerators are used, due to the possibility of destroying healthy cells surrounding the organ in treatment [19].

In more modern times, the magneto-induction of hyperthermia is one of the commonly studied therapies for treating cancer, just like its advancement in its most fragile points. In simplified form, this technique consists of exposing tumor tissue to an alternating magnetic field. For magnetism, hyperthermia uses a magnetic field to place the ferromagnetic α-Fe_2O_3 nanoparticles within the target region, keeping the field within a determined frequency for a specific length of time. Currently, most of the nanoparticles used in hyperthermia are superparamagnetic particles as long as these can generate more heat in smaller magnetic fields than ferromagnetic nanoparticles [20].

The magnetic field is not absorbed by living tissues and can be applied in deep regions of the body. When magnetic particles are subjected to a varying magnetic field, some heat is generated due to hysteresis loss. The amount of heat generated depends on the nature of the magnetic material and the parameters of the magnetic field established [20].

This heat manifests itself in the immediate surrounding diseased tissue where, if temperatures can be maintained above the therapeutic threshold of 42 °C for 30 minutes or more, the cancer is destroyed. Cancer cells are destroyed at temperatures above 43 °C, while normal cells can survive at higher temperatures. The heat can be generated by applying an appropriate magnetic field. The size of the crystals is submicrometric and these materials are not only biocompatible but also bioactive, which could be useful for bone tumors [21]. Many studies with magnetic particles for hyperthermia have been made in order to establish a therapeutic effect on various types of tumors through experiments with animals [22], or with cancer cell cultures [23].

3.2.3 Contrast for Magnetic Resonance Imaging

Magnetic resonance imaging (MRI) is a technique that has the potential to scan the human body based on the principle of magnetic resonance. Thus, to improve the quality of the images of these scans, research is being done to develop new contrast agents for nuclear magnetic resonance imaging. This way the use of these new agents (magnetic fluids) is becoming routine and usually administered orally. However, there are limitations to the available agents due to toxicity, tissue absorption and their effectiveness.

As is known some research groups propose the use of açaí as a contrast agent, due to the effects of paramagnetic ions such as iron, cobalt and manganese contained in the fruit pulp. With the first *in vivo* studies, there have already been good results with regard to the uniformity of the signal, besides the benefits of the ingestion of a natural fruit with practically no toxic effects on the human body [19].

For the use of contrast, there should be a clinically reliable process, i.e., there is need for study and development of clinical instruments used to measure the desired properties, thus ensuring the quality of the material in question. The route of the material also needs to be known, i.e. the path from intake to disposal, to see how the ingested particles are eliminated by the body. For this it is necessary to be able to measure in terms of three-dimensional location, concentration and magnetic stability [19].

Recently, developments in magnetic resonance have allowed *in vivo* images with microscopic resolution. The visualization and control of stem cells is also possible through magnetic resonance by tagging cells magnetically [24]. Magnetic nanoparticles have been used for *in vivo* imaging through magnetic resonance imaging (MRI) as contrast agents for molecular and cellular imaging [25,26]. Superparamegnetic magnetite is used in these procedures as an agent that differentiates between healthy and pathological tissues. However, these superparamagnetic particles are usually coated with a layer of polysaccharides to ensure colloidal stability [27].

The most commonly used contrast agent consisting of superparamagnetic, iron oxide nanoparticles, is Feridex IV, manufactured by Advanced Magnetics Inc [28] which is used to scan both the liver and some vesicular disturbance. The magnetic particles coated with polymers have been successfully developed by Song and colleagues [29]. Along with the features described above, there is interest in making these MRI agents able to provide enough heat to destroy malignant tissue, besides their function as contrast agent, whenever a low frequency alternating magnetic field is applied to them. Consequently, based on this same material composed of magnetic nanoparticles and a modified MR machine, alternating fields would be used, not only for diagnosis but also for therapy of the patient in one single session, facilitating treatment.

3.2.4 *Ferrofluids*

The so-called magnetic fluids or liquids occur in the form of colloidal suspension, which are attracted in a magnetic field generated by magnets. This suspension, called magnetic fluid or ferrofluid, comprises suspended magnetic particles of about 10 nm in a suitable liquid carrier [30]. Due to thermal agitation, the particles remain suspended since a repulsive force occurs between them that keeps the distance between them. Such suspensions should be stable over time when it comes to superparamagnetic nanoparticles, even when subjected to magnetic forces [19]. The principle of operation of ferrofluids in human tissues occurs based on a magnet with a considerably strong magnetic field placed near the tumor. The magnetic nanoparticles are injected into the vein and attracted by the Earth's magnetic field. When they arrive at the tumor they are absorbed by the tumor cells and the drug is administered. The substance that surrounds the nucleus has an affinity with the tissue of interest. The nanoparticles are coupled to antibodies, which are the only cells capable of recognizing tumors. The tumor cells have different receptors from healthy ones. The antibodies attach themselves to the tumor receptors and are absorbed by the tumor. And with them the nanoparticles carry the drug.

The use of magnetite in ferrofluids was originally proposed for high performance seals in space applications. Ferrofluids require superparamagnetic, nanometer-sized particles dispersed in aqueous or organic mediums. A ferrofluid has no magnetic moment, except when under the influence of an applied field. An external magnet is, therefore, able to trap the fluid in a specific location to act as a seal. Ferrofluids are currently used in computer disk sealing units, and in vibrating environments in place of conventional seals. Another application of magnetic fluids is given by the containment of oil spills at sea with the use magnetic barriers. It can also be used for sealing cracks in tanks containing potentially hazardous materials, the separation of various materials (waste and oil), for positioning abdominal organs during surgery and the development of smart shock absorbers [19].

3.3 Magnetic Storage Media

Formerly very fine, iron oxide particles, sometimes with the addition of cobalt, were used as permanent magnets of recording discs. Today, most computer disks are made of metal films, usually cobalt-based alloys, less than a 100 nm thick [31]. Currently, new technologies have emerged such as active heads, which generally are based on the change of electrical resistivity of some materials in the presence of magnetic fields (magneto resistance). Since it was recently noted that this technology is much more potent in certain nanomaterials than in conventional materials, this property was named "giant magneto resistance". Among such materials are complex structures of very thin films and granular materials consisting of nanoscale magnetic particles immersed in base metals (Cu, Au and Ag) [30].

When data storage is concerned, the particles should be stable, the switchable magnetic state should not be affected by temperature fluctuations. For optimum performance in recording, the particles must have inherent characteristics: high coercivity, high remanence, resistance to corrosion, friction and temperature changes. Maghemite is useful in data recording and storage applications because of its physical and chemical stability. Often it is doped or coated with cobalt (1-5%) area in order to improve its coercivity and its storage capacity [30].

The coated nanoparticles have higher thermal stability than their doped counterparts and have uniaxial magnetic anisotropy. As a result, the particles of cobalt-modified maghemite are predominantly intended for use in video tapes, audio tapes and magnetic disks. Maghemite nanoparticles embedded in a nonmagnetic matrix also exhibit giant magnetoresistance, which is a decrease in resistance due to an applied magnetic field, and is useful in magnetic recording heads and sensor elements in magnetometers. With the continuous progress of technology and market requirements, the demand has increased for further miniaturization of memory devices, and along with it, the increase of information that can be stored on a hard disk. Put differently, the aim is to increase storage capacity (amount of information stored per unit area). The goal is to further reduce the bit size, i.e., to achieve the nanometer scale, to further increase the density of information storage capacity [32].

A promising study for the concept of magnetic recording and reading of data is based on the control of some properties of the electron. This has happened because of the discovery of "giant magneto resistance", which controls another property, the spin. With this, several models have emerged based on the properties of electrical currents with electrons that have only one direction of spin. This new technology is known as "electron spin", or "spintronics". There are already models of transistors and even commercial non-volatile memories using this technology [30].

3.3.1 Giant Magnetoresistance and Giant Magneto Impedance

Giant magnetoresistance (GMR) and giant magneto impedance (GMI) have been observed and studied intensively during the last decade [1,33]. One of the most important new phenomena is giant magneto resistance, observed in multilayers of certain magnetic films interspersed with non-magnetic metal films. For certain thicknesses of the non-magnetic films, in the order of 1 nm, the resistance of the system varies greatly with the magnetic field applied on it. This phenomenon was discovered in 1989, with Brazilian Professor Mario Baibich as the principal author of the original research [1]. This effect allows the manufacture of a magnetic sensor of very small physical dimensions, which when crossed by an electric current develops a voltage that depends on the magnetic field.

Recently, magnetic reading technology has been revolutionized by the introduction of magneto-resistive heads based on the giant magnetoresistance effect. Technological advances in this area are impressive. The capacity of magnetic recording disks of computers in 1995, for example, was 1 Gigabits/inch². With the introduction of giant magnetoresistant reading heads this capacity rose to 20 Gigabits/inch² in 2002, allowing the manufacture of disk drives with capacities over 100 Gigabits [34].

Many industrial and engineering applications of GMI sensors have been proposed and implemented to date, including computer disk heads, rotating heads, pin-hole detectors, the displacement of detection sensors, navigation direction sensors (electronic compass), field sensors, biomedical sensors, traffic monitoring, antitheft systems and so on. Basically, when a soft ferromagnetic conductor undergoes a small alternating current (AC), a large change in impedance of the conductor can be achieved by applying a magnetic field. This is known as the giant magneto impedance [35].

3.4 Separation of Particles by Magnetic Action

The separation and selection process of specific molecules contained in large volumes of solution represents a big problem for the biological sciences. The use of conventional chromatographic columns can be time consuming and it is in this field that the use of magnetic or magnetizable adsorbents gains importance. In this procedure, the magnetic adsorbent is added to a solution or suspension with the target. This target binds to the adsorbent and the complex is collected from the suspension by a suitable magnetic sensor. The separation process can be speeded up to 35 times. Moreover, another advantage of using magnetic nanoparticles instead of magnetic microparticles is that they allow the preparation of suspensions that are stable in relation to sedimentation in the absence of an applied magnetic field. Magnetic modifications of standard, immunoassay techniques can be valuable for the determination of many biologically active compounds and of xenobiotics, allowing greater speed and more reproducibility. The magnetic labeling of cells and their subsequent isolation has many possible applications in the biomedical field. Especially important among these processes are: 1) detection and removal of circulating tumor cells using the immunomagnetic procedure; 2) the selective separation of CD34 + (stem cells) which opens new possibilities for transplantation of stem cells and genetic manipulation of the hematopoietic system [2,19].

The technique can also be applied to the selection of apoptotic cells, genetically transformed cells or cellular organelles such as lysosome, plasma membrane, etc. Biosensors for pollutants that are not yet possible with current technology provide another target of nanobiotechnology. Applied to public health, these biosensors could lead to the detection of bacterial contaminants in food and water, finding better ways to detect low levels of toxins or provide faster laboratory diagnostics.

3.4.1 *Magnetic Separation for Water Purification*

The magnetic properties present in some classes of materials also aid in the purification of water, since they affect the physical properties of the contaminants in the water. Furthermore, the combination with other processes allows for an improvised, efficient purification technology. The technique used for purifying water acts primarily by magnetic separation. Recently, Yavuz et al. analyzed the magnetic separation technique in various industries: bleaching clay, steel mills and power plants, enrichment of minerals, food industry, water treatment and metal removal [36].

High gradient magnetic separation (HGMS) is a term commonly used in magnetic separation processes [37]. HGMS device consists basically of an electromagnet (magnetised column), as shown in Figure 3.2. When a magnetic field is applied across the column, it produces large field gradients around the magnet, and this attracts the magnetic particles to the surface, capturing them.

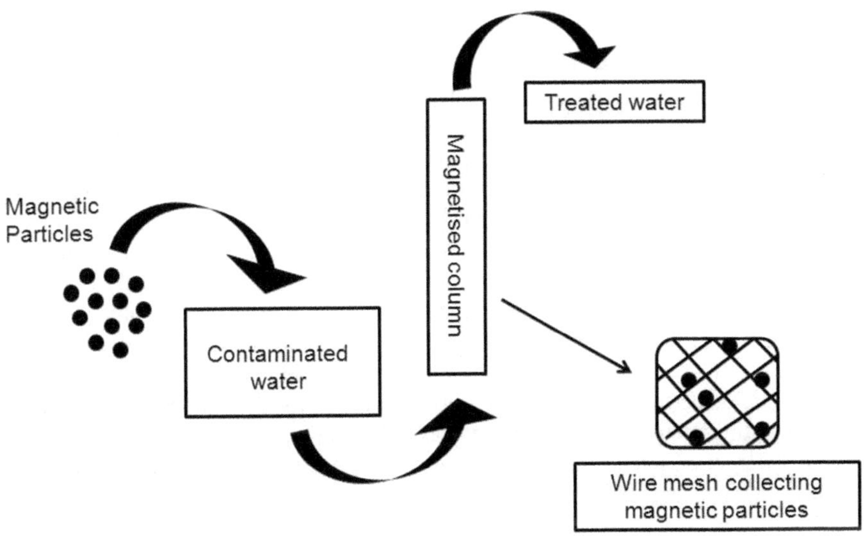

Fig. 3.2. Representation of the process of high gradient magnetic separation (HGMS). Adapted from [34].

3.5 Nanosensors and Devices

As mentioned earlier, the most important aspect for technological applications in the field of magnetic materials is the combination of unusual magnetic properties with reduced particle size. In particular, the following properties are more suitable for applications in sensors and magnetic devices [38,39,40]:

- Absorption of high frequency electromagnetic waves (1-20 GHz);
- Use in codifying magnetic sensors;

- High electrical resistivity and the existence of insulating coating, allowing use in high-precision resistors;
- Giant magneto impedance (up to 600%), allowing use in magnetic field sensors;
- Ferromagnetic resonance for a range of microwave frequencies, allowing use in anti-thief sensors;
- High magnetic permeability at high frequency allowing the application in magnetic shielding;
- High mechanical resistance, allowing the incorporation of microwires inside tissues of the human body;
- Magnetoelastic properties, allowing their use in magnetoelastic sensors.

Electromagnets use electric current to generate magnetic force, and are essential devices for converting electrical energy into mechanical energy. When the current flows through the coils, part of the electrical energy is converted into heat due to resistance of the coils. The resistance of electromagnets not only wastes energy but also causes the generation of heat, which is harmful in many applications.

The parts produced with magnetic materials for automotive applications can be divided into two groups: those that convert motion into an electrical signal, and those that convert an electrical signal into motion. In the first case, the material works under constant magnetization, and requires only moderate induction. Simple iron or phosphorus iron, of medium or high density is commonly used, along with ferritic stainless steel, which combines soft magnetic properties with good corrosion resistance and stretching [41].

Magnetic nanosensors can act as biomarkers whose function is to warn about diseases like cancer, while they are still in their early stages. This is likely to become much easier thanks to a biosensor chip developed by researchers at Stanford University. The sensor is up to 1.000 times more sensitive than any technology currently in clinical use. It is precise regardless of which body fluid is being analyzed. This biosensor can detect proteins in a range of concentrations three times higher than any other existing method, the researchers say. The nanosensor chip can also search up to 64 different proteins simultaneously and has proved effective in early detection of tumors in mice, suggesting that it can detect the disease early [42].

There are also biosensors of pollutants which are not possible with current technology and therefore constitute a target of future nanotechnology. Applied to public health, the biosensors could lead to the detection of bacterial contaminants in food, to find better ways to detect low levels of toxins or providing faster laboratory diagnostics [2,19].

3.5.1 Automotive Applications

For automotive applications, the sensors are needed because they have good accuracy, high functionality and reliable operation under adverse environmental conditions: temperatures from – 40 °C to 150 °C, temperature shock, humidity, fog or even the use of engine oil. And furthermore, the sensors can withstand the mechanical vibrations and the acceleration values can be up to 200 times the acceleration of gravity. An example of magnetic sensors in the production of Robert Bosch GmbH is the LWS3 wheel sensor, developed for the electronic stability program (ESP), which prevents vehicles from twirling. The absolute measurement covers a range of 1560 ° (4.3 turns) with a resolution of 0.1°. The principle of operation acts on a steering column that drives two gear wheels with magnets. The angular position of the gear wheels is measured, and due to different numbers of teeth, the gears move at different speeds. The combination of both measures of angular positions allows for the calculation of the angle of the wheel in the range of 4.3 turns, and a mathematical algorithm allows for the improvement of the accuracy, the correction of mistakes and to verify the functioning [43].

Another interesting application of sensor devices can be found in automobile brakes. The basic operating principle of the Anti-lock Brake System (ABS) is a sensor with rings, these rings are toothed and are coupled to the axle that rotates with the wheel. The movement of the teeth causes an oscillating current in the sensor, which frequency is associated with the speed of the wheel. This way, when a brake of the wheel occurs [44], the oscillating current is interrupted. There are many different designs, with different profiles of teeth or with teeth located on the face or outside diameter of the sensor, but all perform essentially the same function. In addition to its magnetic properties, the ABS sensor rings should be sufficiently elastic so as not to break during assembly on the axle of the wheel, and it also needs to have good corrosion resistance. Ferritic stainless steels (ferritic stainless steel) are suitable materials to provide this combination of properties, for example.

3.6 Catalysts and Pigments

The association of magnetite and hematite as catalysts can be used in a variety of industrially important reactions, including the synthesis of NH_3 (Haber process); reactions that include the dehydrogenation of ethylbenzene into styrene, the oxidation of alcohols, as well as the large-scale development of butadiene. Magnetite and hematite are the semiconductors and can catalyze oxidation / reduction reactions. Hematite has also been used as support material for gold in catalysts for the oxidation of carbon monoxide at low temperatures [45]. Maghemite, carbon and magnetite composites, on the other hand, were obtained to reduce the amount of undesirable N_2 in fuel oil [46].

Iron oxides can be used as acid/base catalysts [47] and to catalyze the degradation of the copolymer acrylnitrileebutadieneestyrene in fuel oil. All three forms of magnetic iron oxides (hematite, maghemite and magnetite) are commonly used in synthetic paint pigments. They have a number of desirable attributes for this application, since they display a range of colors with pure tones and high hue resistance. They are also extremely stable and highly resistant to acids and alkalis. Hematite-based pigments are red, those based on maghemite are brown, and magnetite pigments are based on black [48]. Yellow-based pigments can be obtained from goethite, which when calcined at 400 °C – 500 °C provides red pigments similar to those obtained with hematite [49]. These pigments are widely used in water-repellents for wood, which allow the wood grain to be seen, while protecting against the harmful effects of sunlight.

Another use of magnetic materials combined with other materials is as adsorbent for removal of contaminants. Carbon is considered an excellent adsorbent: when certain chemicals pass next to the carbon nanostructures, they are fused to the surface and are trapped. However, the initial market for nanoparticles coated in carbon is based on its use as catalysts, or materials used to accelerate chemical reactions. In powder form, it may already be used on an industrial scale, but it is necessary to use a magnet to remove it from the reaction bed [45,48].

3.7 Concluding Remarks

Nanomagnetism includes the artificial structuring of magnetic materials with dimensions below the micrometer, i.e. nanometer and natural occurring magnetic entities such as molecules and clusters. Nanometer-sized magnetic particles are situated at the frontier between classical and quantum magnetism.

It is well established that the magnetic nanomaterials have found promising use in various applications fields, including biomedicine, magnetic storage media, magnetic particle separation, sensors, devices, catalysts, and pigments.

These technological applications represent the breadth and importance of magnetic nano-sized materials and their impact on our everyday lives. Each of the cases provides new and unique challenges that have been investigated with outstanding level of achievement of research scientists.

The magnetic nanoparticles have great potential in applications for innumerable fields of technology. They are already being used successfully in production of nanocomposite magnets, magnetic fluids, and magnetic liposomes.

Many other potential applications could be mentioned as a switch agent for blood flow during cardiac surgery, radioisotopes for use in brachytherapy, the role of mechanical forces on signal transduction, regulation of cellular functions, including cell growth, proliferation, protein synthesis and gene expression, which undoubtedly will lead to wider use of nanoparticle systems in the medical field.

Hence, it is noteworthy that there are perspective discussion questions concerning nanomagnetic materials (nanopowders, nanowires, thin films and magnetic fluids) and their new applications. In addition to technological relevance, other exciting basic science questions remain to be answered, which arise from the interplay of finite size and shape-induced modifications of magnetic behavior combined with interactions across interfaces between magnetically different materials.

The challenge for the future is to apply the nanomagnetism phenomenon observed creatively and their materials to the large-scale industry.

References

[1] Baibich MN, Broto JM, Fert A et al (1988) Giant Magnetoresistance of (001)Fe/(001)Cr Magnetic Superlattices. Physical Review Letters 61:2472-2474

[2] Laçava ZGM, Morais PC (2004) Aplicações biomédicas de nanopartículas magnéticas. Parcerias Estratégica/CGEE 18:73-85

[3] Bader SD (2006) Opportunities in nanomagnetism. Modern Physics 78:1-5

[4] Knobel M (2000) Os superpoderes dos nanomagnetos. Ciência Hoje 159:33-37
[5] Morup S, Tronc E (2004) Superparamagnetic relaxation of weakly interacting particles. Physical Review Letters 72:3278-3281
[6] Dorman JL, Fiorani D, Tronc E (1999) On the models for interparticle interactions in nanoparticle assemblies: comparison with experimental results. Advances in Chemical Physics 202:251-267
[7] Mornet S, Vasseur S, Grasset F, Veverka P et al (2004) Magnetic nanoparticle design for medical applications. Journal of Materials Chemistry 14:2161-2175
[8] Duran N, Mattodo LHC, Morais PC (2006) Nanotecnologia: introdução, preparação e caracterização de nanomateriais e exemplos de aplicação. Artliber. São Paulo
[9] Azevedo MMM (2003) O que é nanobiotecnologia? Atualidades e perspectivas. Instituto de Química. Monograph. Unicamp
[10] Chomoucka J, Drbohlavova J, Huska D, Adam V, Kizek R, Hubalek J (2010) Magnetic nanoparticles and targeted drug delivering. Pharmacological Research 62: 144-149
[11] Pankhurst QA, Connolly J, Jones SK, Dobson J (2003) Applications of magnetic nanoparticles in biomedicine. Journal of Physics D: Applied Physics 36:R167-R181
[12] Drbohlavova J, Hrdy R, Adam V, Kizek R, Schneeweiss O, Hubalek J (2009) Preparation and properties of various magnetic nanoparticles. Sensors 9:2352-2362
[13] Tucek J, Zboril R, Petridis D (2006) Maghemite nanoparticles by view of Mossbauer spectroscopy. Journal for Nanoscience and Nanotechnology 6:926-947
[14] Neuberger T, Schopf B, Hofmann H, Hofmann M, von Rechenberg B (2005) Journal of Magnetism and Magnetic Materials 293:483-496
[15] Faraji A, Wipf P (2009) Nanoparticles in cellular drug delivery. Bioorganic & Medicinal Chemistry 17:2950-2962
[16] Singh R, Lillard JW (2009) Nanoparticle-based targeted drug delivery. Experimental and Molecular Pathology 86:215-223
[17] Reshmi G, Kumar PM, Malathi M (2009) Preparation, characterization and dielectric studies on carbonyl iron/cellulose acetate hydrogen phthalate core/shell nanoparticles for drug delivery applications. International Journal of Pharmaceutics 365:131-135
[18] Arias JL, Lopez-Viota M, Ruiz MA, Lopez-Viota J, Delgado AV (2007) Development of carbonyl iron/ethylcellulose core/shell nanoparticles for biomedical applications. International Journal of Pharmaceutics 339:237-245
[19] Knobel M, Goya GF (2004). Ferramentas magnéticas na escala do átomo. Scientific American Brasil, 58-66
[20] Heintz ELH (2004) Surface biological modification and cellular interactions of magneticspinel ferrite nanoparticles. Thesis. Georgia Institute Technology, Georgia.
[21] Gordon RT, Hines JR, Gordon D (2009) Intracellular hyperthermia - A biophysical approach to cancer treatment via intracellular temperature and biophysical alterations. Medical Hypotheses 5:83-102
[22] Luderer AA, Borrelli NF, Panzarino JN, Mansfield GR, Hess DM, Brown JL, Barnett EH (1983) Glass-ceramic-mediated, magneticfield- induced localized hyperthermia: response of a murine mammary carcinoma. Radiation Research 94:190-198
[23] Chan DCF, Kirpotin DB, Bunn Jr PA (1993) Synthesis and evaluation of colloidal magnetic iron oxides for the site specific radiofrequency- induced hyperthermia of cancer. Journal of Magnetism and Magnetic Materials 122:374-378
[24] Fawell S, Seery J, Daikh Y, Moore C, Chen LL, Pepinsky B, Barsoum J (1994) Tat-mediated delivery of heterologous proteins into cells. Proceedings of the National Academy of Sciences of the United States of America 91:664-668

[25] Ai H, Flask C, Weinberg B, Shuai X, Pagel MD, Farrell D, Duerk J, Gao JM (2005) Magnetite-Loaded Polymeric Micelles as Ultrasensitive Magnetic-Resonance Probes. Advanced Materials 17:1949-1952

[26] Duanmu C, Saha I, Zheng Y, Goodson BM, Gao Y (2006) Dendron-functionalized superparamagnetic nanoparticles with switchable solubility in organic or aqueous media: Matrices for homogenous catalysis and potential MRI contrast agents. Chemistry of Materials 18:5973-5981

[27] Babes L, Denizot B, Tanguy G, Le Jeune JJ, Jallet P (1999) Synthesis of Iron Oxide Nanoparticles Used as MRI Contrast Agents: A Parametric Study. Journal of Colloid and Interface Science 212:474-482

[28] http://www.berleximaging.com/html/feridex/index.html Accessed May 2010

[29] Song H.T, Choi JS, Huh YM, Kim S, Jun YW, Suh JS, Cheon J (2005) Surface Modulation of Magnetic Nanocrystals in the Development of Highly Efficient Magnetic Resonance Probes for Intracellular Labeling. Journal of the American Chemical Society 127:9992-9993

[30] Holm C, Weis JJ (2005) The structure of ferrofluids: A status report. Current Opinion in Colloid & Interface Science 10:133-140

[31] Knobel M (2000) Partículas finas: superparamagnetismo e magnetorresistência gigante. Revista Brasileira de Ensino de Física 22:387-395

[32] Herzer G, Vazquez M, Knobel M, Zhukov A, Reininger T, Davies HA, Grossinger R, Sanchez JL (2005) Present and future applications of nanocrystalline magnetic materials. Journal of Magnetism and Magnetic Materials 294:252-266

[33] Mohri K, Uchiyama T, Shen LP, Cai CM, Panina LV (2001) Sensitive micro magnetic sensor family utilizing magneto-impedance (MI) and stress-impedance (SI) effects for intelligent measurements and controls. Sensors Actuators A 91:85-90

[34] http://www.comciencia.br/reportagens/nanotecnologia/nano14.htm Accessed July 2010

[35] Phan MH, Peng HX (2008) Giant magnetoimpedance materials: Fundamentals and application Progress in Materials Science 53:323–420

[36] Yavuz CT, Prakash A, Mayo JT, Colvina VL (2009) Magnetic separations: From steel plants to biotechnology. Chemical Engineering Science 64:2510-2521

[37] Ambashta RD, Sillanpää M (2010) Water purification using magnetic assistance: A review. Journal of Hazardous Materials 180:38-49

[38] Chiriac H, Ovari TA (1996) Amorphous glass-covered magnetic wires: Preparation, properties, applications. Prog Mater 40:333-407

[39] Zhukov A (2002) Glass-coated magnetic microwires for technical applications. Journal of Magnetism and Magnetic Materials 216:242-245

[40] Zhukov A, Blanco JM, Zhukov A, González JJ (2001) J Phys D: Appl Phys 34: L113-L116

[41] Bas JA, Calero JA, Dougan MJ (2003)Sintered soft magnetic materials. Properties and applications. Journal of Magnetism and Magnetic Materials 254–255:391–398

[42] http://news.stanford.edu/pr/2009/pr-detect-cancer-101309.html Accessed June 2010

[43] Treutle CPO (2001) Magnetic sensors for automotive applications. Sensors and Actuators A: Physical 91:2-6

[44] McHenry M E, Laughlin D E (2000) Nano-scale materials development for future magnetic applications. Acta materialia 48:223-238

[45] Hutchings GJ, Hall MS, Carley AF, Landon P, Solsona BE, Kiely CJ, Herzing A, Makkee M, Moulijn JA, Overweg A, Fierro-Gonzalez JC, Guzman J, Gates BC (2006). Role of gold cations in the oxidation of carbon monoxide catalyzed by iron oxide-supported gold. Journal of Catalysis 242:71-81
[46] Brebu M, Uddin MA, Muto A, Sakata Y, Vasile C (2001) The effect of PVC and/or PET on thermal degradation of polymer mixtures containing brominated ABS. Energy & Fuels 15:559-564
[47] Shi F, Tse MK, Pohl MM, Bruckner A, Zhang SM, Beller M (2007) Tuning catalytic activity between homogeneous and heterogeneous catalysis: Improved activity and selectivity of free nano-Fe2O3 in selective oxidations. Angewandte Chemie International Edition 46:8866-8868
[48] Lam UT, Mammucari R, Suzuki K, Foster NR (2008). Processing of Iron Oxide Nanoparticles by Supercritical Fluids. Industrial & Engineering Chemistry Research 47:599-614
[49] Gaedcke H (1998) Iron oxide pigments. In: Buxbaum G (ed). Industrial Inorganic Pigments, VCH, Weinheim

Abbreviations

ABS - anti-lock brake system
AC - alternating current
DNA - deoxyribonucleic acid
GMI - giant magneto impedance
GMR - giant magneto resistance
HGMS - high gradient magnetic separation
MRI - magnetic resonance imaging

4 Optoeletronic and Ferroeletric Applications

Mônica Jung de Andrade[1], Felipe Fernandes de Oliveira[1], Biana Faraco[1], Renato Bonadiman[2], and Vânia Caldas Sousa[1]

[1] Departamento de Engenharia de Materiais, Universidade Federal do Rio Grande do Sul, 90035190, Porto Alegre, Brazil
[2] Instituto Nokia de Tecnologia, Manaus, Brazil
E-mail: mja0612@gmail.com

Abstract. In this chapter, some of the tendencies in the fields of nanostructured materials for optoeletronic and ferroelectric applications in terms of research and industry are presented. Special emphasis is given to carbon nanotubes, graphene, perovskite, aurivillius and thin films. In the field of nanotechnology applied to ceramic materials, there are different methods of fabrication and processing of these materials, such as powder form, bulk or thin films. The appropriate choice of processing depends on the application desired. The main tendency for both optoeletronic and ferroelectric applications is the use of thin films, which can contribute to the minituarization of devices such as transistors, photovoltaics, liquid crystals, light emitting diodes and sensors. Different techniques can be used to produce these nanometric structures as thin films, including in-situ growth and post-synthesis techniques. The selection of the adequate technique depends on the material and thickness of the film desired, which will have direct effect over the properties.

Keywords: carbon nanotubes, graphene, perovskite, aurivillius, thin films.

4.1 Optoelectronic Applications

Optoelectronic technology has been enabling a revolution in the fields of communication, energy and medicine. The range of products varies from TV remotes to powerful telescopes. For instance, the Brewer Company, Inc. has drawn attention as materials provider to compound semiconductors, Microelectromechanical systems (MEMS), and to the optoelectronics industries. The company's products include anti-reflective coatings, wafer bonding materials, etch protective coatings, and coat and bake equipment for clean rooms and planarization tools [1]. Together with Pixelligent, an emerging leader in developing nanocomposites for demanding applications in the electronics and military markets, it will develop nanocomposites which are expected to have a wide range of semiconductor and microelectronics applications such as coatings that can create brighter and more efficient light emitting diode (LED) displays and new materials that can improve the performance of Blu-ray discs [2].

Electronic materials based on organic chemistries are attractive options for a number of important electronics applications, for instance, radio-frequency identification (RFID), photovoltaics, sensors and displays. These materials are frequently classified as organic electronics. However, a more accurate designation would be organic optoelectronics, since light is often involved in practical implementations of this technology and the related materials. In this sense, one direction for developers of materials of organic electronics materials is to develop new materials that combine the advantages of organic and inorganic materials. For example, new formulations combine organic materials with carbon nanotubes (CNTs). These hybrids will offer the printability, wide area capabilities and suitability for flexible substrates of organic materials with the high performance of CNTs. As a result carbon nanotube electronics firms, such as Nantero, are now looking to organic electronics as promising markets [3].

One promising application area is related to printed electronics. Printable technology may have a promising future when electronic devices are concerned. Through the dot arrangement of functional inks, it is possible to obtain flexible devices with innovative applications [4]. A wide variety of materials can be inkjetted, e.g. polymers, adhesives, solders, and nano-particles [5]. Silver, gold and copper nano-particle-based inks are most common. However, CNT-based inks are already in an advanced development stage in order to become commercially available. Companies such as Chisso Corporation [6], NanoMas Technologies [7] and Advanced Nano Products [8] can be mentioned as ink developing companies.

Recent developments indicate that this technology has a potential to be a disruptive technology that could replace actual electronics industry standard processes and reach a market bigger than the actual silicon industry. In this sense, researches for flexible electronics devices (FEDs) are rapidly increasing worldwide [9]. Since the market potential is huge, the optoelectronics industry has also interest in this technology and some applications are already being studied.

Sensors are a huge application area and therefore also include a number of industries (aerospace, automotive, food supply, medical, etc.). There are, however, still few companies involved with ink applications in order to obtain sensors. An example is Nanoident Company [10] which is one of the world leaders in printed electronic sensors and develops optoelectronic sensors. There is, therefore, an opportunity to develop inks of organic materials and, consequently, to obtain sensors though printable electronics technology. These chemical sensor materials can be integrated with MEMS creating, for instance, environmental sensors. Not only just environmental sensors could be created, however, but also biosensors, gas sensors and also photovoltaic cells. These cells could be integrated to flexible electronics providing the energy source needed to solve one of the big issues of future flexible mobiles and electronics, being an efficient and flexible energy source.

The Clean Technology International Corp. has developed Carbon NanoSphere Chains (CNSC) which are totally new nanocarbon materials. This CNSC Material is catalyst-free and 100% carbon as produced, with no embedded extraneous material. Hydrogen atoms are removed from the nanomaterial as part of the production process, and hydrogen gas is a byproduct. The electrical conductivity of Clean Technology International materials is excellent, with specific resistance levels determined by the degree of functionalization. Current research is revealing the unique properties of this material, indicating applications in a variety of industries including electronics, composites, airframe coatings, protective armor, fuel additives, super capacitors, hydrogen fuel cells [11].

All these optoelectronic applications such as displays, touch screens, light-emitting diodes and solar cells require materials with low sheet resistance R_s and high transparency. In this context, CNT and graphene arise as attractive replacements for indium-tin-oxide (ITO) and other transparent conductors. These nanomaterials can add a level of operational flexibility that is not possible with current transparent conductors and rigid glass substrates.

4.1.1 Carbon Nanotube

Carbon nanotubes (CNTs) are allotropes of carbon with a cylindrical nanostructure and considered as nearly one-dimensional structures according to their aspect ratio (length usually in micrometers over diameter of generally 1 - 20 nanometers) [12]. CNTs have unique physical and chemical properties that yielded more than 60000 publications in the last 19 years (according to Science Citation Index, ISI [13]). CNT technology offers many potential opportunities for the assembly of optoeletronics devices [14,15,16,17].

Many researchers focus their work in the field of optoeletronics using CNTs as pure films [18,19,20] or in the form of composites [21]. LEDs, carbon nanotube field-effect transistors (CNFETs), flexible transparent electrodes, next-generation displays and next-generation flat-panel screens, known as field-emission displays (FED), are some examples.

Many companies followed this trend, including Toray Industries Inc. [22], Motorola [23], Unidym Inc. [24] and Eikos Inc. [25]. The Japanese Toray [22], e.g., intends to expand its variety of products by using its CNTs technology. In turn, Motorola [23] has already developed flat screen displays with CNT technology. The American-based Unidym Inc. [24] has developed transparent conductive films with CNT-based technology that enable applications in touch panels, liquid crystal displays (LCDs) and thin film solar markets. The CNT ink called Invisicon^{TM} from Eikos Inc. [25], another American-based company, also enables the manufacture of transparent conductive coatings over different substrates for displays and photovoltaic cells applications, for example.

For all these applications, usually the production of CNTs films is needed, which can be produced by direct growth [26,27] or post-growth [28-36] methods.

In the case of direct growth techniques, it is possible to obtain aligned (by gas flow, electric field or substrate structure) networks composed of long (several millimetres) CNTs, but the substrate should be compatible with the temperatures of synthesis of CNTs and some by products can be deleterious for the transmittance and electrical conductivity of the film. Besides the long aspect ratio of CNTs obtained by direct growth methods, the transmittance *versus* surface conductance of films are usually very similar to those obtained by post-growth techniques.

Post-growth techniques such as spray-coating (or air-brushing) [28,29,30], spin-coating [31,32], dip-coating [28,30,33], vacuum filtration method [28,30,34,35] and electrophoretic deposition [28,30,36] enable the production of films over several substrates and using purified CNTs. For some applications, like solar cells for example, the roughness of the CNT networks (CNTNs) can play an important role because, in thin layered configurations, unpacked CNTs would cause a short circuit [37]. The average thickness of solar cells is in the order of 200 nm and CNTs are generally micrometer long, so unpacked CNTNs might cause a short circuit if a metallic CNT reaches the other electrode. In general, electrophoretic deposition (ED) provides films composed by randomly oriented CNTs with the smoothest surface among other post-growth techniques [28,30] and this is probably due to the good packing of the CNTs over the electrode [36]. This method generates films with similar surface conductance versus transparency in comparison to other post-growth methods with randomly oriented CNTs, but it would be possible to increase transparency by dissolving the thin oxide layer formed during the electrophoresis process that causes light scattering [36]. Among these techniques, dip-coating (DC) is an example of a technique that provides smooth films composed by partially (each dipping decreases alignment) oriented CNTs (Figure 4.1), which allow increasing surface conductance in a given transparency [28,30,33]. These results show that the performance of the SWCNT networks through the DC method has reached parameters which may already make them suitable for applications such as solar cells, flat panel displays, touch screens and electrostatic dissipation [28,30].

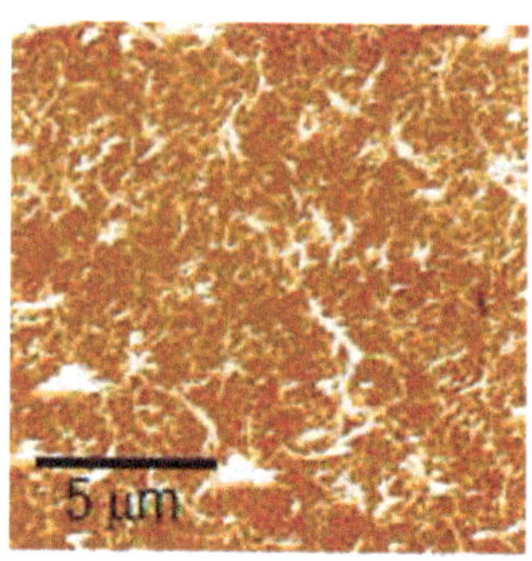

Fig. 4.1. Microstructure of a partially aligned CNTN obtained by DC [30].

4.1.2 Graphene

Graphene was first produced by micromechanical exfoliation of graphite [38] and since that work this two-dimensional monatomic sheet has attracted the interest of several researchers [39]. Not surprisingly the groundbreaking achievements on graphene were crowned with a Nobel Prize in physics [40]. Graphene has a combination of high mobility (fast carrier transport) and optical transparency (higher transparency over a wider wavelength range than single-walled carbon nanotube films, thin metallic films and ITO [41]), in addition to flexibility, robustness and environmental stability. These interesting features pushed graphene-based electronics and optoeletronics to become one of the most important research topics in nanomaterial science.

Its unique optoeletronic properties also attracted potential investors from industry such as from IBM [42], contributing to push research to large-scale production of graphene. One of the most promising techniques is chemical vapor deposition (CVD) on transition metal substrates [43,44,45,46], since it enables the production of large area, transparent, and highly conducting graphene films, of a single to a few-layers of graphene. In this technique, metal foils can be used for the substrate, thus making it compatible with roll-to-roll manufacturing [45]. Even though it is just a few years since graphene was first produced, it is already possible to find graphene for commercialization [47,48,49]. Figure 4.2 displays a graphene film produced through CVD over a copper foil. As demonstrated in the literature [50], the placement of graphene layers on arbitrary substrates is also possible (transfer method).

Fig. 4.2. Microstructure of an annealed Cu foil with large Cu grains with single layer CVDGraphene™ film coverage [47].

A collaboration of Asian research groups, e.g., recently reported the production of 30-in graphene films on copper foil, with sheet resistance of approximately 125 ohms/square and optical transmittance of 97.4% [45]. They tested the films in a touch screen, where the film survived 6% strain, double that of ITO.

Graphene films were also tested as anodes for application in photovoltaic devices and a power conversion efficiency (PCE) of up to 1.71% was demonstrated (which is 55.2% of the PCE of a control device based on ITO) [46]. Researchers from the Max-Planck Institute successfully demonstrated the application of graphene films as window electrodes in solid-state dye-sensitized solar cells, but it also showed a relatively lower efficiency than metal oxide coating-based solar cells [51].

In turn, there are also reports of organic light emitting diodes (OLEDs) based on graphene with a performance similar to a control device based on conventional ITO transparent anodes [52]. Besides this, LCDs might be graphene's first realistic commercial application [53].

4.2 Ferroeletric Applications

Since the 50s, the effect of particle size on the ferroelectric properties has been studied. The main effect occurs with the ferroelectric – paraelectric transition of the material under study [54].

Understanding the behavior displayed by ferroeletric materials from micrometric dimensions until a few nanometers is quite complicated. Importantly, the physical boundary conditions are different for processed materials such as particles, grains in sintered ceramics, fibers and thin films [54]. This is generated in the processing characteristics such as the degree of distortion of the crystal lattice, the mechanical tension between film and substrate, among other mechanisms of defect formation, promoting changes in the thermal, mechanical, electrical and optical properties of the material.

Consequently, the system properties are strongly correlated with material interfaces. Such contours can be either ceramic electrodes, grain boundaries in ceramics interdomain interactions. Moreover, the study of thin films becomes complex due to the high dependence of the thickness and the substrate on the dielectric and electrical properties [62,74].

For nanostructured ferroelectric materials, it is important to determine the critical size, since the processing of materials with smaller sizes favors the disappearance of ferroelectric behavior. For lead zirconate titanate (PZT) nanodisks, the critical size is equal to 3.2 nm [55]. It is clear that nanostructures around this size are difficult to process. Thus, techniques for material synthesis and combustion synthesis [56,57], sol-gel [58], the hydrothermal method [59] are required to obtain nanometric particles.

The behavior of nano-sized particles found in ferroelectrics has attracted great interest in order to study the phenomenon known as polarization swirl (vortex-to-polarization). This phenomenon occurs due to the reorientation of dipoles in a specific geometry, forming a toroid moment [60,61]. Therefore, materials such as nanoislands [62], nanodisks [55], nanodots [63] and nanotubes [64] have become the subject of research for developing new technologies based on this phenomenon, which enhances the ferroelectric and dielectric properties of nanomaterials [55].

4.2.1 *Perovskite (ABO_3)*

A material used to get plenty of Perovskite is PZT. It consists of a solid solution between lead titanate (PT) and lead zirconate (PZ), allowing different formulations in stoichiometry. The compound ferroelectric PZT presents a variety of properties determined by its chemical composition and with the addition of lanthanum, electro-optic characteristics are obtained. The different compositions are easily processed via solid state reaction, where their precursors (oxides) are mixed and subsequently heat treated in search of a specific phase [65].

The PZT system with the addition of lanthanum disturbs the crystal lattice of PZT, lanthanum replaces the site of lead (site's network type ABO_3). Therefore, its inclusion on the site promotes vacancies A and B of the unit cell. Moreover, a decrease of Curie temperature (Tc) and other ferroelectric and piezoelectric properties are affected, this will be further discussed later in the item [66].

For electro-optic applications, it is necessary to obtain transparent materials, with materials that are becoming increasingly dense. The technique of hot pressing helps to close the pores and increases transparency for ferroelectric electro-optical applications. This technique consists in performing the constant heating of a powder compaction during the sintering process. Moreover, the technique also helps to inhibit the growth of grains, facilitating the attainment of ceramics with nanograins [67].

However, there are advantages in manufacturing ferroelectric thin films such as PLZT, with nanometric thicknesses, like offering lower voltages in electrical circuits, the largest economy in the use of raw materials, miniaturization of the developed devices. In addition, it yields a decrease in densification temperature of hundreds of degrees compared to the conventional process (through solid state reaction) [68].

Today, there are many new devices in addition to conventional applications of ferroelectric materials such as PZT-based piezoelectric materials. Even a well-studied material in the field of nanoscience has new properties, such as in the field of solar energy. Meng Qin et al. [69] observed mechanisms in photovoltaic thin films and found that the effects of interface and the electrodes drastically affect its efficiency. An increase in the dielectric constant of the electrodes substantially improves the photoconductive properties of PZT. Moreover, an efficiency of 19.5% is theoretically possible for ferroelectric ultrathin films or nanostructures.

4.2.2 *Structure Layer (Aurivillius)*

The family of layered bismuth compounds, known as Aurivillius, presents itself as a great alternative not only for not using lead, but also for having superior properties with respect to the polarization retention of PZT, which may be replaced by it in non-volatile ferroelectric random access memory (NVFRAM) devices. Other potential applications of these materials are dielectric, piezoelectric and pyroelectric materials [68].

In the field of nanotechnology applied to ceramic materials, we should note that applications are becoming more concrete, with the possibility of fabrication and processing of materials in powder form [57], bulk [58,59] and thin films [70]. The most appropriate choice of geometry depends on the type of application desired.

The various applications of nanoscale materials allow a good absorption by the industry responsible for manufacturing and marketing such materials. The class of powdered materials seeks to take advantage of the high surface area in processes such as photocatalysis [56,71]. The materials in bulk are developed in large quantities by research groups that investigate the influence of compositional variations and doping with new elements. Materials with this geometry minimize interference, differently when a substrate for thin films or nanomaterials in large areas is used.

Figure 4.3 shows a structure of a nano-powder with a bismuth layered structure, more specifically with $SrBi_2Ta_2O_9$ composition, showing morphology with sizes ranging from 20 nm to 50 nm. Nanostructured particles present high surface area and they become feasible in such fields of study as photocatalysis. According to Yingxuan et al. [71], $SrBi_2Ta_2O_9$ has applications in the process of the photocatalytic decomposition of water.

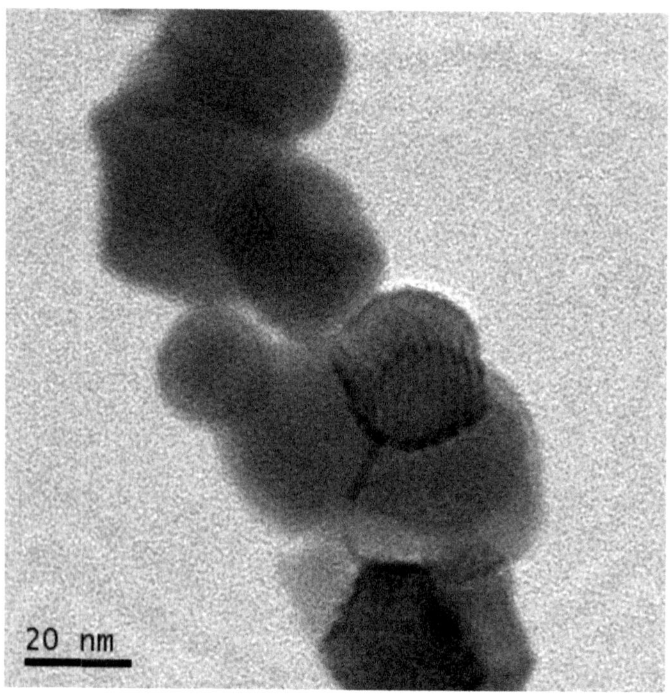

Fig. 4.3. Image of transmission electron microscopy (TEM) of the nanostrutured ferroelectric powder $SrBi_2Ta_2O_9$ with cristalline structure of bismuth layers (Aurivillius) [72].

The vacancy formation occurs inherently with increasing temperature, this provides energy to the atoms causing vibration in the atomic lattice, allowing the ions to jump elsewhere. The increase in vacancies, both through intrinsic and extrinsic factors, increases the ionic conductivity of these materials when subjected to heating due to thermally activated behavior. Figure 4.4 shows the Arrhenius curve for

$SrBi_2Ta_2O_9$. In this figure, one can observe that the temperature increases from right to left, showing an increase in electrical conductivity resulting from the mechanisms of ion conduction.

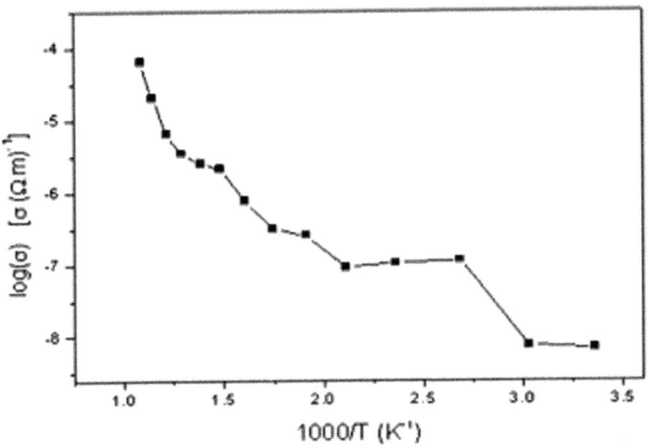

Fig. 4.4. Arrhenius curve for electrical conductivity as a function of the temperature of the sintered ceramic $SrBi_2Ta_2O_9$ (relative density = 83%) [72].

4.2.3 *Thin Films*

Commercial development in this area was initiated with piezoelectric crystals for applications such as transducers, specifically in sonar, during the 2nd World War, when there was a major driving force for the advancement of this research.

Non-volatile memories using ferroelectric materials were initially proposed in 1952, using a crystal of barium titanate ($BaTiO_3$) with a thickness of 300 micrometers. Companies like IBM and Ford have invested in a pioneering way, thus giving rise to a technological breakthrough in the commercialization of ferroelectric materials [73].

However, the use of single crystalline materials leads to high costs, making them virtually unviable. Thus, using polycrystalline materials technology becomes increasingly necessary. Recently, ferroelectric thin films have been studied extensively in the manufacture of electronic devices, as shown in Figure 4.5. Lead zirconate titanate (PZT) and strontium bismuth tantalate (SBT) are good candidates for thin films in this type of device, where low energy consumption, reduced floor space and larger data storage capacity make it the lightest and most functional NVFRAMs [67,68].

One advantage of ferroelectric memories is the low energy expenditure when compared with magnetic memories. Moreover, the access time to write or read information in the areas allocated to ferroelectrics is of the order of nanoseconds. Commercially, companies such as Fujitsu, Samsung, Matsushita have access times of 100 ns, 60 ns and 60ns, respectively. Products such as those from Fujitsu have 32 Kb of memory, which may seem low, but are employed, for example, in the Playstation 2 [74].

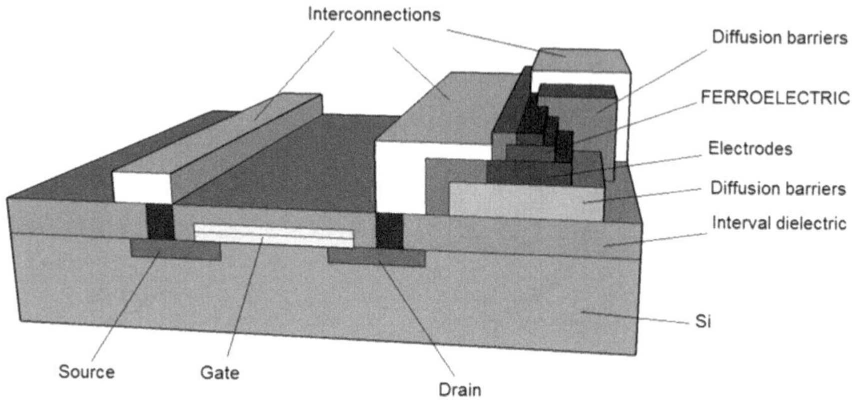

Fig. 4.5. Schematic diagram of NVFRAM memory device.

Ceramic thin films are used in various technological applications such as coatings for corrosion protection, optical applications, electronic devices, sensors, thermal barriers, etc. There are several techniques for deposition of thin films, subdivided into physical and chemical methods.

The deposition techniques such as laser ablation [75], sputtering [76] are some of the physical method techniques. The chemical method employs techniques such as DC [77], spin-coating [78], metal-organic chemical vapor deposition (MOCVD) [79], spray pyrolysis [80], etc. Moreover, the method requires a chemical precursor solution to obtain the films. Table 4.1 presents the main techniques of deposition and their respective fields in industry or in scientific research and technology.

The spray pyrolysis technique is a simple and low cost technique to obtain ceramic thin films with a thickness from a few micrometers down to nanometric sizes. In addition, it is possible to change the parameters of the technique to produce nanostructured powders [80]. A disadvantage of this technique is the need for a solution of precursor ions.

Table 4.1. Some techniques to deposit thin films.

Method	Advantage	Disadvantage	Field of application
Sputtering [76]	CMOS compatible	Large capital expenditure, difficult to control with complex oxides	Semiconductor industry, research and development
Spray pyrolysis [80]	Inexpensive	Difficult to control the thickness for deposition on large areas, requires a chemical precursor solution	Solar cells, sensors and optoelectronics
CVD [79]	CMOS compatible	Large capital expenditure	Semiconductor industry
Dip coating [77]	Inexpensive	High cost of precursors, the thickness gradient on the films.	Research and development

4.3 Concluding Remarks

According to the literature, carbon-based nanostructures have demonstrated great potential for applications in optoeletronics in the last two decades and will probably be the base of next generation malleable optoeletronics. This will enable different designs and higher versatility than conventional optoeletronics devices. To produce displays, touch screens, light-emitting diodes and solar cells with carbon-based nanostructures, it is necessary to produce thin films, which can be produced by in-situ growth (CVD) or post-synthesis techniques (spray-coating, spin-coating, dip-coating, vacuum filtration method, electrophoretic deposition and transfer method). Light and malleable optoeletronics that can be produced with these technologies have many advantages, such as devices that can be easy to carry and are adjustable to clothes or that minimize the breakage of solar panels during their installation due to the fragility of the currently used materials (ITO), for example.

The wide range of properties offered by ferroelectric materials makes them indispensable for nanotechnology. Potential applications can be visualized in the areas of electronics with increasingly cheap and portable sensors and actuators. Furthermore, photovoltaic properties observed in ferroelectric materials make them interesting for alternative energy research. An area with so many possibilities both from engineering and scientific perspective offers students a new field of application of nanoscale engineering. It is important to emphasize that the proper selection of the method of power synthesis and film deposition are important to control the characteristics of the particles, like size and morphology, which have direct effect in the obtainment of the films or bulk. This control will be decisive in the final properties of the material studied.

Thus, the main tendency for both optoeletronic and ferroelectric applications is the use of these nanometric structures as thin films, which can contribute to the minituarization of devices like transistors, photovoltaics, liquid crystals, light emitting diodes and sensors.

References

[1] http://investing.businessweek.com/research/stocks/private/snapshot.asp?privcapId=4422135 Accessed October 2010
[2] http://www.nanotech-now.com/news.cgi?story_id=36423 Accessed October 2010
[3] http://www.nantero.com Accessed October 2010
[4] Bonadiman R, Salazar MMP, Silva OC (2009) Reliability of inkjet printed PEDOT:PPS conductive traces over Polyimide substrates. In: The Polymer Processing Society, Europe/Africa Regional Meeting.
[5] Kawamura Y et al (2006) LTCC Multilevel Interconnection Substrate with Ink-jet Printing and Thick Film Printing for High.-Density Packaging. In: Proc. of 39th International Symposium on Microelectronics. pp. 462-69
[6] http://www.chisso.co.jp Accessed October 2010
[7] www.nanomastech.com/ Accessed October 2010
[8] www.anapro.com/ Acessed October 2010

[9] Bonadiman R and Salazar MMP (2010) Reliability of Inkjet Printed Ag Traces Submitted to High Current Density Test. In: Proceedings of the Annual Technical Conference of the Society of Plastics Engineers. pg. 1072-1076
[10] http://www.nanoindent.eu/default.asp Accessed October 2010
[11] http://www.cleantechnano.com/applications.html Accessed October 2010
[12] O'Connell, M.J. (2006). Carbon Nanotubes – Properties and Applications. New York, USA: Taylor & Francis.
[13] http://sub3.isiknowledge.com Accessed October 2010
[14] Avouris P, Chen ZH, Perebeinos V (2007) Carbon-based electronics. Nature nanotechnology 2:605-615
[15] Avouris P, Freitag M, Perebeinos V (2008) Carbon-nanotube photonics and optoelectronics. Nature Photonics 2:341-350
[16] Avouris P (2004) Carbon nanotube electronics and optoelectronics. MRS Bulletin 29:403-410
[17] Li Y, Qian F, Xiang J, Lieber CM (2006) Nanowire electronic and optoelectronic devices. Materials Today 9:18-27
[18] Fanchini G, Unalan HE, Chhowalla M (2006) Optoelectronic Properties of Transparent and Conducting Single Wall Carbon Nanotube Thin Films. Appl Phys Lett 88:191919.1-191919.3
[19] Star A, Lu Y, Bradley K, Grüner G (2004) Nanotube Optoeletronic Memory Devices. Nano Lett 4:1587-1591
[20] Saran N, Parikh K, Suh DS, Munoz E, Kolla H, Manohar SK (2004) Fabrication and characterization of thin films of single-walled carbon nanotube bundles on flexible plastic substrates. Journal of the American Chemical Society 126:4462-4463
[21] Boussaad S, Harmer MA (2006) Single-walled carbon nanotube composites. US Patent Application 2006/0240238 A1
[22] http://www.toray.com/technology/index.html Accessed May 2010
[23] http://mediacenter.motorola.com Accessed May 2010
[24] http://www.unidym.com Accessed May 2010
[25] http://www.eikos.com Accessed May 2010
[26] Zhang YG, Chang A, Cao J, Wang Q, Kim W, Li Y, Morris N, Yenilmez E, Kong J, Dai H (2001) Electric-field-directed growth of aligned single-walled carbon nanotubes. Applied Physics Letters 79:3155-3157
[27] Lima MD, De Andrade MJ, Skákalová V, Nobre F, Bergmann CP, Roth S (2007) In-situ synthesis of transparent and conductive carbon nanotube networks. Physica Status Solidi (RRL) - Rapid Research Letters 1:165-167
[28] De Andrade MJ, Lima MD, Skakalova V, Bergmann CP, Roth S (2007) Electrical properties of transparent carbon nanotube networks prepared through different techniques. Physica Status Solidi (RRL) - Rapid Research Letters 1:178-180
[29] Kaempgen M, Duesberg GS, Roth S (2005) Transparent carbon nanotube coatings. Appl Surf Sci 252:425-429
[30] Jung De Andrade M (2010) Study of Electrical Properties of 2- and 3-Dimensional Carbon Nanotubes Networks. Thesis, Universidade Federal do Rio Grande do Sul and Université Paul Sabatier
[31] Star A, Lu Y, Bradley K, Grüner G (2004) Nanotube optoelectronic memory devices. Nano Lett 4:1587-1591
[32] Castro MS (2007) Alternative Conductive Coatings Based on Multi-Walled Carbon Nanotubes. Shaker Verlag, Saarbrücken

[33] Ng MHA, Hartadi LT, Tan H, Poa CHP (2008) Efficient coating of transparent and conductive carbon nanotube thin films on plastic substrates. Nanotechnology 19:205703.1-205703.5
[34] Wu ZC, Chen ZH, Du X, Logan JM, Sippel J, Nikolou M, Kamaras K, Reynolds JR, Tanner DB, Hebard AF, Rinzler AG (2004) Transparent, conductive carbon nanotube films. Science 305:1273-1276
[35] Rowell MW, Topinka MA, McGehee MD, Prall HJ, Dennler G, Sariciftci NS, Hu LB, Grüner G (2006) Organic solar cells with carbon nanotube network electrodes. Appl Phys Lett 88:233506.1-233506.3
[36] Lima MD, De Andrade MJ, Bergmann CP, Roth S (2008) Thin, conductive, carbon nanotube networks over transparent substrates by electrophoretic deposition. J Mater Chem 18:776-779
[37] Tenent RC, Barnes TM, Bergeson JD, Ferguson AJ, To B, Gedvilas LM, Heben MJ, Blackburn JL (2009) Advanced Materials 21:3210-3216
[38] Novoselov KS; Geim AK; Morozov SV, Jiang D, Zhang Y, Dubonos SV, Grigorieva IV, Firsov AA (2004) Electric field effect in atomically thin carbon films. Science 306:666-669
[39] Gem AK (2009) Graphene: Status and Prospects. Science 324:1530-1534
[40] http://nobelprize.org/nobel_prizes/physics/laureates/2010/index.html Accessed october 2010
[41] Bonaccorso F, Sun Z, Hasan T, Ferrari AC (2010) Graphene photonics and optoelectronics. Nature Photonics 4:611-622
[42] http://www.ibm.com/Search/?sn=23&q=GRAPHENE&v=16&us=utf&lang=en&cc=us&Search=Search Accessed October 2010
[43] Li XS, Cai WW, An JH, Kim S, Nah J, Yang DX, Piner R, Velamakanni A, Jung I, Tutuc E, Banerjee SK, Colombo L, Ruoff RS (2009) Large-Area Synthesis of High-Quality and Uniform Graphene Films on Copper Foils. Science 324:1312-1314
[44] Kim KS, Zhao Y, Jang H, Lee SY, Kim JM, Kim KS, Ahn JH, Kim P, Choi JY, Hong BH (2009) Large-scale pattern growth of graphene films for stretchable transparent electrodes. Nature 457:706-710
[45] Bae S, Kim H, Lee Y, Xu XF, Park JS, Zheng Y, Balakrishnan J, Lei T, Kim HR, Song YI, Kim YJ, Kim KS, Ozyilmaz B, Ahn JH, Hong BH, Iijima S (2010) Roll-to-roll production of 30-inch graphene films for transparent electrodes. Nature Nanotechnology 5:574-578
[46] Wang Y, Chen X, Zhong Y, Zhu F, Loh KP (2009) Large area, continuous, few-layered graphene as anodes in organic photovoltaic devices. Applied Physics Letters 95:063302.1-063302.3
[47] http://www.firstnano.com/applications/cvdgraphene/ Accessed October 2010
[48] https://graphene-supermarket.com/home.php Accessed October 2010
[49] http://www.graphenelab.com/ Accessed October 2010
[50] Reina A, Son H, Jiao L, Fan B, Dresselhaus MS, Liu ZF, Kong J (2008) Transferring and Identification of Single- and Few-Layer Graphene on Arbitrary Substrates. J Phys Chem C 112:17741–17744
[51] Wang X, Zhi L, Müllen K (2007) Transparent, Conductive Graphene Electrodes for Dye-Sensitized Solar Cells. Nano Lett 8: 323–327
[52] Wu J, Agrawal M, Becerril HA, Bao Z, Liu Z, Chen Y, Peumans P (2010) Organic Light-Emitting Diodes on Solution-Processed Graphene Transparent Electrodes. ACS Nano 4:43–48

[53] Blake P, Brimicombe PD, Nair RR, Booth TJ, Jiang D, Schedin F, Ponomarenko LA, Morozov SV, Gleeson HF, Hill EW, Geim AK, Novoselov KS (2008) Graphene-Based Liquid Crystal Device. Nano Lett 8:1704–1708
[54] Zhao Z, Buscaglia V, Viviani M, Buscaglia MT, Mitoseriu L, Testino A, Nygren M, Johnson M, Nanni P (2004) Grain-size effects on the ferroelectric behavior of dense nanocrystalline $BaTiO_3$ ceramics. Physical Review B 70:024107.1-024107.8
[55] Naumov II, Bellaiche L, Fu H (2004) Unusual phase transitions in ferroelectric nano-disks and nanorods. Nature 432:737-740
[56] Civera A, Pavese M, Saracco G, Specchia V (2003) Combustion synthesis of perovskite-type catalysts for natural gas combustion. Catalysis Today 83:199-211
[57] Patil KC, Aruna ST, Mimani T (2002) Combustion Synthesis. Current Opinion in Solid State and Materials Science 6:507-512
[58] Jain R, Gupta V, Mansingh A, Sreenivas K (2004) Ferroelectric and Piezoelectric properties of non-stoichiometric Sr1-xBi2+2x/3Ta2O9 ceramics prepared from sol-gel derived powders. Materials Science & Engineering B 112:54-58
[59] Dias A, Moreira RL (2007) Production of Sr-deficient bismuth tantalates from microwave-hydrothermal derived precursors: structural and dielectric properties. Journal of Physics and Chemistry of Solids 68:645-649
[60] Naumov I, Fu H (2007) Vortex-to-polarization phase transformation path in ferro-electric $PB(ZrTi)O_3$ nanoparticles. Physical Review Letters 98:077603.1-077603.4
[61] Naumov II, Bellaiche LM, Prosandeev SA, Ponomareva IV, Kornev IA (2009) Ferro-electric nanostrucuture having switchable multi-stable vortex states. US 7593250 B2
[62] Kim Y, Han H, Kim Y, Lee W, Alexe M, Baik S, Kim JK (2010) Ultrahigh density array of epitaxial ferroelectric nanoislands on conducting substrates. Nano Letters 10:2141-2146
[63] Schilling A, Byrne D, Catalan G, Webber KG, Genenko A, Wu GS, Scott JF, Gregg JM (2009) Domains in ferroelectric nanodots. Nano Letters 9:3359-3364
[64] Morrison FD, Luo Y, Szafraniak I, Nagarajan V, Wehrspohn RB, Steinhart M, Wendorff JH, Zakharov ND, Mishina ED, Vorotilov KA, Sigov AS, Nakabayashi S, Alexe M, Ramesh R, Scott JF (2003) Ferroelectric Nanotubes. Review on Advanced Materials Science 4:114-122
[65] Haertling GH (1999) Ferroelectric Ceramics: History and Technology. Journal of the American Ceramic Society 88:797-818
[66] Mesquita A (2007) Preparação e caracterização de materiais ferroelétricos de composição $Pb_{1-x}La_xTiO_3$ em escala nanométrica. Dissertation, University of São Paulo.
[67] Moulson AJ, Hebert JM (2003) Electroceramics: Materials, Properties, Aplications. Second edition. John Wiley & Sons Ltd., New York
[68] Nalwa HS (2002) Handbook of thin film materials – vol. 3: Ferroelectric and Dielectric thin films. Academic Press, San Diego
[69] Qin M, Yao K, Liang YC (2009) Photovoltaic mechanisms in ferroelectric thin films with the effects of the electrodes and interfaces. Applied Physics Letters 95:022912.1-022912.3
[70] Lu T, Cao W (2002) Generalized continuum theory for ferroelectric thin films. Physical Review B 66: 024102.1-024102.5
[71] Yingxuan L, Gang C, Hongjie Z, Zhonghua L, Jingxue S (2008) Electronic structure and photocatalytic properties of $ABi_2Ta_2O_9$ (A=Ca, Sr, Ba). Journal of Solid State Chemistry 181:2653-2659

[72] Oliveira FF (2010) Síntese por combustão de pós $SrBi_2Ta_2O_9$, produção de corpos cerâmicos e sua caracterização quanto à microestrutura e propriedades elétricas. Dissertation, Universidade Federal do Rio Grande do Sul

[73] Daglish M, Kemmitt T (2000) Ferroelectric thin films – research, development and commercialisation. IPENZ Transactions 27:21-24

[74] Dawber M, Rabe KM, Scott JF (2005) Physics of thin-film ferroelectric oxides. Reviews of Modern Physics 77:1083–1130

[75] Kidoh H, Ogawa T, Morimoto A, Shimizu T (1991) Ferroelectric properties of lead zirconate titanate films prepared by laser ablation. Applied Physics Letters 58: 2910-2912

[76] Haccart T, Cattan E, Remiens D (2002) Dielectric, ferroelectric and piezoelectric properties of sputtered PZT thin films on Si substrates: influence of film thickness and orientation. Semiconductor Physics, Quantum Eletronics & Optoeletronics 5:78-88

[77] Brinker CJ, Hurd AJ (1994) Fundamentals of sol-gel dip coating. Journal de Physique III 4:1231-1242

[78] Kato K, Zheng C, Finder JM, Dey SK (1998) Sol-gel route to ferroelectric layer-structured perovskite $SrBi_2Ta_2O_9$ and $SrBi_2Nb_2O_9$ thin films. Journal of the American Ceramic Society 81:1869-1875

[79] Kijima T (1999) New low temperature preparation of ferroelectric $Bi_4Ti_3O_{12}$ thin film by MOCVD method. Applications of Ferroelectrics 26:93-101

[80] Perednis D (2003) Thin film deposition by spray pyrolysis and the application in solid oxide fuel cells. Thesis, Swiss Federal Institute of Techonolgy Zurich

Abbreviations

CMOS - Complementary metal-oxide-semiconductor
CNFETs - carbon nanotube field-effect transistors
CNSC - Carbon NanoSphere Chains
CNTs - Carbon nanotubes
CNTNs - Carbon nanotubes networks
CVD - chemical vapor deposition
DC - dip-coating
ED - electrophoretic deposition
FED - field-emission display
ITO - indium-tin-oxide
LCDs - liquid crystal displays
LEDs - Light emitting diodes
MEMS - Microelectromechanical systems
MOCVD - metal-organic chemical vapor deposition
NVFRAM - non-volatile ferroelectric random access memory
OLEDs - organic light emitting diodes
PCE - power conversion efficiency
PLZT - lead lanthanum zirconate titanate
PT - lead titanate
PZT - lead zirconate titanate
RFID - Radio-frequency identification
SBT - strontium bismuth tantalite
TEM - transmission electron microscopy

5 Nanostructured Materials for Energy Applications

Cibele Melo Halmenschlager[1], Mônica Jung de Andrade[1], Diego Pereira Tarragó[2], and Célia de Fraga Malfatti[2]

[1] Departamento de Engenharia de Materiais, Universidade Federal do Rio Grande do Sul, 90035190, Porto Alegre, Brazil
[2] Departamento de Engenharia Metalúrgica, Universidade Federal do Rio Grande do Sul, 91501970, Porto Alegre, Brazil
E-mail: celia.malfatti@ufrgs.br

Abstract. There is great interest in research related to alternative forms of electricity production in order to promote increased quantity and quality of the energy system, maintaining and enhancing environmental sustainability, economic, emphasizing efficient use of renewable energy resources. In this context, it is important to develop new technologies for energy generation, especially those from renewable resources. Nanostructured materials have been extensively researched in application like lithium ion batteries, supercapacitors, solar cells and fuel cells. This chapter contains a brief overview of some companies that are already dealing with nanostructured materials for lithium ion batteries, supercapacitors, solar cells and fuel cells, followed by recent developments on research of nanostructured materials for elements of Intermediate Temperature Solid Oxide Fuel Cell.

Keywords: lithium ion batteries, supercapacitors, solar cells and fuel cells.

5.1 Companies Engaged in the Development of Nanostructured Materials for Batteries, Supercapacitors, Solar Cells and Fuel Cells

Nanostructured materials have been extensively researched in application like *lithium ion batteries*, *supercapacitors*, *solar cells* and *fuel cells*.

Companies like the north-american A123 Systems [1], the british Nanotecture [2] and the japanese Toyota [3] are extensively researching new nanostructures for devices in Lithium ion batteries. Lithium-ion batteries used in cell phones and PCs, and in cordless power tools are proving the technology to power hybrid and electric vehicles.

A123Systems is one of the world's leading suppliers of high-power lithium ion batteries and their nano *Li-Ion battery* is based on Nanophosphate™ [1] technology developed at the Massachusetts Institute of Technology. Their batteries are claimed to charge to "high capacity" in 5 minutes and to offer 4 times higher thermal conductivity than conventional Lithium-Ion cylindrical cells. This technology would be used by the *plug-in electric hybrid car* from General Motors, the Chevrolet Volt. This *nanophosphate technology* is also being used by Valence Technology Inc. [4] together with different active materials than the traditionally used for *Li-Ion batteries* (Saphion Li-Ion technology).

Nanostructured *lithium titanate spinel oxide* has been another alternative recently investigated by AltairNano to obtain a one-minute recharge *nanobattery* [5]. Another nanomaterial that has been extensively researched in this field is carbon nanotubes (CNTs), as exemplified by the *carbon nanotube battery* developed by Next Alternative Inc [6] that would be extend the recharge life by at least 4 times with the new *CNT lead/lead-acid battery* (typically, a lead-acid battery presents 200 cycles).

One of the investments of Toyota has lately been nanostructured *lithium-ion batteries*, more specifically nanocomposite cathodes of lithium iron/manganese phosphates (with olivine structure) reinforced with carbonaceous materials that enabled large active material discharge capacity with high capacity retention rate as 84% approximately after 500 cycles [7], and anodes composed by lithium-alloying nanoparticles that are able to minimize overall volume expansion for are some of the investments of in research [8]. One practical example of the investments in the field of batteries is the Plug-In Hybrid Electric Vehicles (PHEVs) with the first-generation *lithium-ion battery* [3].

Researchers at the University of California (Los Angeles, USA) have successfully developed a "carbon nanotube ink" for manufacturing flexible batteries that conduct more efficiently than a conventional battery using printed electronics techniques [9]. The resultant *zinc-carbon battery*, with less than a millimeter thick, is composed by carbon cathode and manganese oxide electrolyte components can be printed as different layers on a surface, over which an anode layer of zinc foil can be printed.

Nanotecture developments on *nanoporous* materials in different forms — thin-film, monolithic, powder, discrete particles — obtained by a technique of liquid crystal templating, enabled the design and production of high performance *batteries and supercapacitors* [2]. This technology to obtain micron-sized particles with defined and ordered pores penetrating all the way through the particle, thus with high-surface-area, was discovered at the University of Southampton. It exploits liquid crystal (surfactants in solution that align themselves) honeycomb structures that act as ideal templates for the alignment and deposition of other materials such as metal salts.

The korean company LG Chem together with Stanford University, Hanyang University and Ulsan National Institute of Science & Technology developed Letter Silicon Nanotube Battery Anodes [10]. They reported that the capacity in a Li-ion full cell consisting of a cathode of $LiCoO_2$ and anode of Si nanotubes demonstrates a 10 times higher capacity than commercially available graphite even after 200 cycles.

The japanese company Hitachi Maxell, one of the leading manufacturer of battery products, developed technology for portable, high-energy-density, long-life, small-scale (direct methanol and polymer electrolyte) fuel cell (DMFC and PEFC, respectively) [11]. This technology is based on the high-activity of catalysts made up of nanoparticles which can be as small as 2 nanometers in size and it allows high oxygen–reduction reaction activity (Figure 5.1).

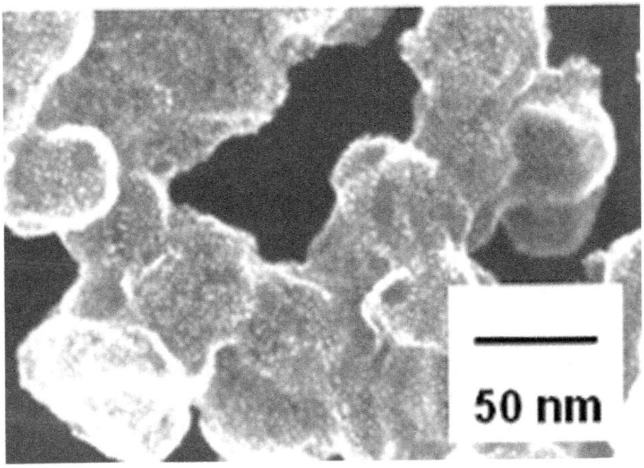

Fig. 5.1. Electron microscope image of platinum-ruthenium-phosphorous catalyst after reduction-separation on acetylene black (the white granular objects are the catalyst particles) [11].

More achievements on nanostructured materials have also enabled to obtain more efficient *supercapacitors* by companies like Nanoener Inc. [12] and Nanotecture [2]. Nanoener Inc. [12] specialty is development of active material deposition technologies that combine the nanoparticles formation from liquid and solid phases (like physical vapor deposition, chemical vapor deposition and other methods of deposition). Their expertise enables the production of thin (10-40 μm) electrodes based on $LiMn_2O_4$, $LiCoO_2$, MnO_2, C, Si with nanoparticles of active material integrated into current collector with for applications that include *fuel cells* and *supercapacitors.*

The benefits of nanotechnology in *solar cells* are (i) reduce costs of materials, processing and instalation (by carbon nanotube technology or thin films) in comparison to traditional ones; (ii) reach higher efficiency levels (by quantum dots technology) than conventional ones. Thus, these achievements in *solar energy* using nanostructured materials attracted companies like Cyrium Technologies [13], Illuminex Corp. [14], Nanosolar [15] and Unidym [16].

Cyrium Technologies [13], e.g., offers solar cells (structure: AlGaInP–Ga(In)As–Ge) with advanced quantum dot nanotechnology (Quantum Dot Enhanced Cell, QDEC) in standard cell size products with efficiency levels of ~40% at >500-1000 suns and a minimum efficiency offering of 38% on a standard 10x10mm cell.

Illuminex Corp. [14] is developing a novel photovoltaic device architecture based on a novel *silicon nanowire processing* technique that enables the production of photovoltaic devices to be constructed on metallic threads for photovoltaic textile applications, foils or applied layers for conformal fitting photovoltaic coatings, or simply on glass sheets which would compete head-to-head with the established and dominant Si wafer-based PV device market. The structure of the nanowire array enables high conversion efficiency devices to be built at low cost. Conformal silicon nanowire photovoltaic coatings could be applied to the body of an electric or hybrid automobile, be used on buildings and houses with a minimal aesthetic impact, integrated into the cases of portable electronic devices, used on military installations and equipments.

Nanosolar [15] developed a generation of efficient and low-cost *solar electricity cells*, which are light-weight, flexible, and easily adjustable in shape based on flexible plastics, solution coating, and self-assembling nanostructures.

Unidym's [16] transparent conductive films based on *carbon nanotubes* could offer a lower cost solution that is far more compatible with high volume production techniques for new generation of *solar cells*. Unidym works with some thin film solar manufacturers to incorporate its CNT technology into various thin film lines. In dye-sensitized cells, CNTs can greatly enhance the conversion efficiency by providing pathways for the current generated by the conversion of sunlight to reach the cell's electrodes, while in organic photovoltaic's, functionalized fullerenes substantially enhance current efficiency.

Fundamental research on nanotechnology is providing fuel cell manufacturers the possibility to make *fuel cells* more durable, reliable and portable with a cost competitive with traditional power sources. For example, three-dimensional nanostructures improve current electrocatalysts (orders of magnitude) that have traditionally been expressed on a flat surface. All these achievements enable to develop small fuel cells that can be used to replace batteries in handheld devices such as PDAs or laptop computers or even replace them in electric cars. Ceramatec [17], Clean Technology International Corp. [18], Nanoener Inc. [12], NexTech [19], Toyota [3], Unidym [16], USNanocorp [20] and Samsung [21,22] are some of the companies that are already using nanostructured materials for *fuel cells*.

Nanocrystalline ceramic powders — single phase or two phase mixtures of Y_2O_3, CaO, TiO_2, SnO_2, Al_2O_3, ZrO_2, ZrO_{2-x} mol% Y_2O_3, Fe_2O_3, and ZnO — for *fuel cells* are being commercialized by Ceramatec [17].

According to Clean Technology International Corp. [18], their solid *Carbon NanoSphere Chains* (CNSC) are nanostructured carbon materials with a spherical morphology and a high electrical conductivity electrical that enhances efficiency and are cost effective with applications in a variety of industries, including *supercapacitors*, *hydrogen fuel cells* and composites for *solar energy*.

Nanoener Inc. [12] activities include the deposition of nanomaterials for the production of super thin (10-40 μm) electrodes based on $LiMn_2O_4$, $LiCoO_2$, MnO_2, C, Si with nanoparticles of active material integrated into current collector with possible applications in *supercapacitors* and *fuel cells*. Their technology is based on high-pressure vapor flow, gas (vapor) flow with particles of initial material in liquid state and gas (vapor) flow with particles of initial material in solid state. This allows them to deposit at high rate (more than 100 μm /sec) ceramic or ceramic-metal coatings of determined porosity on metal or ceramic substrates.

NexTech [19] investments in *fuel cells* include: fuel cells that converts hydrocarbon fuels to electric power, fuel cell stacks in the 100-W to 5-kW power range, high performance fuel cells including both tubular and planar cells for portable, stationary, and transportation applications, catalysts for fuel processing. Their innovations include *nanocrystalline electrolytes* like Gadolinium/Samarium Doped Ceria or Yttria-Stabilized Zirconia nanopowders.

Toyota [3] concerns on sustainable mobility and alternative-fuel vehicle technologies investments portfolio also includes polymer electrolyte *fuel cells*, as demonstrated by the Toyota (Hydrogen) Fuel Cell Hybrid Vehicle, FCHV (and FCHV-adv). Their systems consist of four hydrogen fuel tanks, an electric motor, a nickel-metal hydride battery, and a power control unit.

Unidym [16] is developing *CNT-based fuel cell electrodes* for powering portable electronics in collaboration with Johnson Matthey and Motorola.

USNanocorp [20] has successfully produced electrolyte and anode for solid oxide fuel cell (SOFC) at intermediate temperatures based on the *nanoparticles* of *Lanthanum Strontium Gallate Magnesite* (LSGM as electrolyte) and *yttrium doped strontium titanate* (SYT as anode) that are obtained by wet chemical synthesis and further atomized to a sprayable form capable of being fed into industrial thermal spray equipment that produces dense LSGM electrolytes and porous SYT anodes in a single sequential thermal spray operation.

Samsung's *fuel cell* system contains Rotating Disk Electrodes with catalyst layers containing *nanoparticles of Pt* (and its alloy) on the carbon carrier (that could be *carbon nanotubes*, *carbon nanofibers*, *carbon nanowires*, *carbon nanohorns, or carbon nanorings*) and other on an inorganic oxide carrier [21,22]. They are also testing a second-generation prototype of the direct methanol *fuel cell* produced by MTI MicroFuel Cells Inc. [23] that decreases cost by eliminating the pumps and controls of the typical water-recirculation increasing performance with reduction of size and costs to viable levels, using a sandwich of nano-engineered hydrophilic and hydrophobic membrane layers that pull in methanol and push back water.

The north-american PolyFuel [24] nanoengineered *hydrocarbon membranes* (based on molecules that self-assemble to a film) specifically optimized for liquid methanol prevents the problem of excess of methanol flowing through the PEM without giving up its hydrogen electrons to generate electricity (thus, reacting instead with oxygen from air to produce excess water that must be removed). Their membranes have been introduced into UK-based Johnson Matthey's manufacturing process, a major *fuel-cell* component supplier.

Germany's Smart Fuel Cell AG [25] has already taken advantage of the growing market of *portable cells* enabled by nanotechnology. Their portable direct methanol fuel cells (under the name EFOY, or, Energy For You) is composed by a *nanoparticle catalyst* and a *nanostructured three-dimensional membrane electrode* assembly (MEA), increasing performance with reduction of size and costs to viable levels.

5.2 Fuel Cell

Nowadays, alternative energies that damage the environment minimally are needed. In this sense, there are researches on dispositives that depend, in their majority, on natural factors such as wind, heat and water flow. A viable alternative are the fuel cells, which do not depend on local factors like other types of alternative energy. A fuel cell is an electrochemical device consisting of an electrolyte, an anode, and a cathode which continuously converts the chemical energy of a fuel directly into electrical energy. Various types of fuel cells are available, including direct methanol fuel cells, alkaline fuel cells (AFC), proton exchange membrane fuel cells (PEMFC), phosphoric acid fuel cells (PAFC), molten carbonate fuel cells (MCFC), and solid oxide fuel cells (SOFCs). With great energy availability and efficiency around 47%, this clean energy production technology is becoming increasingly more attractive. The high efficiency of these devices occurs due to the production of electricity directly from the chemical reactions promoted by the fuel cells [26,27]. These reactions vary according to the catalysts and to the electrolyte used in the fuel cell and its operating temperature, which are the main factors that determine the cells efficiencies and fuels that can be used in their supply [28]. They affect the reactions' kinetics and the ionic transportation in the cell [29]. Table 5.1 presents the fuel cells named after their electrolytes with some of other characteristics of the different types of fuel cells.

Table 5.1. Fuel cells types and its characteristics adapted from Kirubakaran [28] and Hoogers [29].

Fuel Cell Type	Electrolyte	Operation Temperature (°C)	Potency / Application
AFC	Potassium hydroxide	50 – 120	5kW / spatial and military devices
PAFC	Orthophosphoric acid	180 – 210	200kW / portable
PEMFC	Sulfonic acid polymer fluoridated	60 – 110	5 – 250Kw / automotive
MCFC	Lithium and potassium carbonates	630 – 650	200kW – MW / stationary
SOFC	Ytria-stabilized zirconia	650 - 1000	2kW – MW / stationary

5.2.1 Intermediate Temperature Solid Oxide Fuel Cell

Among these cells, stands the solid oxide fuel cell (SOFC). Great improvements in the materials used in this type of cell resulted in level of technological refinement [30]. However, the large scale dissemination of these materials still depends on some factors like lowering the costs of the materials, a more reliable sealing, the probability of a flaw caused by the high operation temperatures, the anode oxidation and complex processing of the ceramic components.

The researches in the area of SOFC started strongly in the 1960s. In 1962, a revolution occurred in the energy research area. Scientists of Westinghouse Electric Corporation (nowadays known as Siemens Westinghouse) have demonstrated for the first time the possibility of production and operation of these cells, called at the time, "solid electrolyte fuel cell" [30]. The SOFCs are classified according to the material in which the whole cell is supported: the first generation cells are known to be supported in the electrolyte, made of ytria-stabilized zirconia. The electrolyte thickness in these cells is in the range of 100 μm -1 mm [31]. The main disadvantages of these cells are their elevated cost and the need of high operation temperatures to overcome the ohmic losses. The second generation of SOFCs uses one of the electrodes as support [32,33]. Again, because both the anode and the cathode are costly ceramic materials, these are also expensive cells. Currently, the scientific community turns its efforts to research third generation of SOFCs, where the cell is supported in the metallic interconnector. This progress has been made by reducing the thickness of the electrolyte [34] and improving the cathode electrolyte interface reaction (i.e. triple phases boundaries, TPB, to internal diffusion, ID, mechanism) [35]. The lower operating temperature authorizes metallic alloys as possible candidates for interconnects [36,37]. Metallic materials have higher electrical and thermal conductivities, are easier to fabricate,

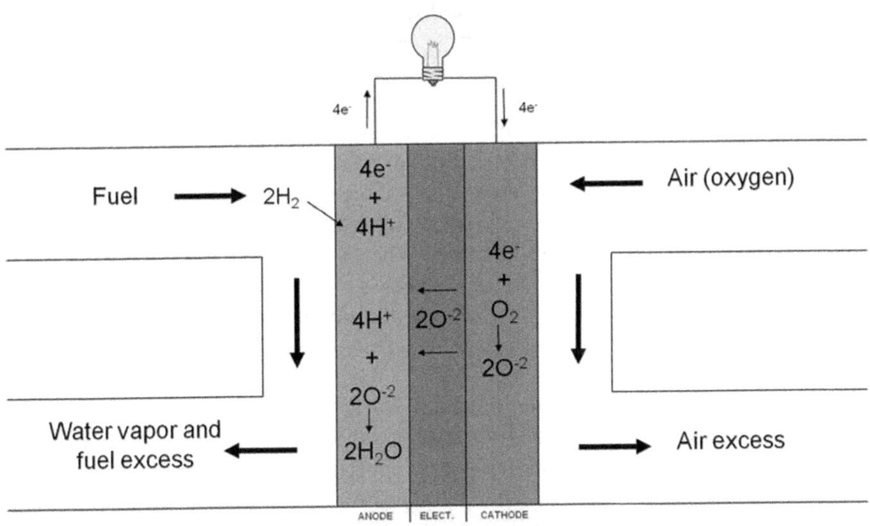

Fig. 5.2. Principle of operation and chemical reactions of a SOFC fuelled with hydrogen.

and, in general, have lower cost compared to the ceramic interconnects [38]. This kind of technology, arises with the advantage of being able to prepare the heart of the cell in extremely thin layers, lowering spends with the materials and also allowing smaller operation temperatures, i.e., the Intermediate Temperature Solid Oxide Fuel Cell (ITSOFC).

The operating principle of SOFCs (Figure 5.2) is to convert chemical energy from fuels (hydrogen, ethanol, gasoline, diesel, etc.) into electrical energy by the natural tendency that the oxygen has to react with hydrogen, causing an explosion with high efficiency and low emissions of pollutants in the atmosphere. Due to the high temperature, materials used as electrodes are already reactive enough to dispense the use of catalysts, which are expensive noble metals, such as platinum, that do not present available reserves to meet a large demand. Another advantage of SOFCs is the high theoretical efficiency and the co-generation (electricity/heat) that allows the internal reforming of the gas, within the fuel cell body. Moreover, as all cell components are solid, an easier fabrication is possible with more flexible configurations. Yet, even with all advances reached so far, there are still some limitations in the use of this technology in industrial scale, like the high temperature of work. This high temperature can activate undesirable interfacial reactions between the components, causing a rapid deterioration of the cell. These reactions limit and difficult the choice of materials used in this technology.

The first Brazilian prototypes of the SOFCs are already in test. These prototypes were built by researches from Institute for Energy and Nuclear Research (known as IPEN). The first was built with electrolyte made of yttria-stabilized zirconia (YSZ). These materials were extracted from minerals and industrially processed. The second cell built and in testing has been developed with the electrolyte-based cerium oxide and the third was developed with a base electrolyte of barium cerate and barium zirconate. These cells have been tested in Japanese institutions and now Brazil enters this race in order to optimize these cells with the main objective of using fuel more flexible in relation cells more classical [39].

State of the Art Electrolyte

Ytria-stabilized zirconia (YSZ) is classically used as SOFC's electrolyte [40,41]. Zirconium oxide can present tree distinct crystalline phases: monoclinic, tetragonal or cubic [42]. The last two, are stable phases when the ceramic is submitted to high temperatures or when other oxides, such as Y_2O_3, CaO, among other rare-earth oxides, are added to the zirconia. When doped with ions with smaller valence, the zirconia has its number of vacancies raised and, owing to that, the ionic conductivity also increases [43]. The incorporation of yttria in the zirconia crystal lattice can be described by a reaction of defects using the notation of Kröger and Vink: $Y_2O_3 \rightarrow 2YZr' + 3O^O{}_x + V_O''$ [44].

Electrolytes with better ionic conductivity then zirconia have been studied to substitute it. LSGM [31,45] has been successfully when applied with a metallic Ni anode, presenting ionic conductivity in temperatures around 400°C. The problem with such material is its low chemical stability. This material reacts with Cr_2O_3 used in the interconnectors besides reacting with the Ni from the anode, limiting the use of this material [31]. Another exhaustly studied material for SOFC's electrolytes is gadolinium-doped ceria (CGO). This material has better ionic conductivity than YSZ and can be used in lower temperatures. However, in 600ºC the reduction $Ce^{+4} \rightarrow Ce^{+3}$ occurs. This reduction confers electric conduction to the material and the cell can suffer a short circuit [31,46].

The greatest challenge nowadays is to reach fuel cells operating around 700 ºC without efficiency losses. The operation temperature can be lowered when the electrolyte thickness is reduced and the densification optimized. Several researches are looking for this objective, the obtaining of thin and dense electrolytes [47].

A lot of techniques have been studied to obtain electrolytes as thin films. Different methods are being tested, such as physical vapor deposition (PVD), sputtering, chemical vapor depositions (CVD), tape casting, spray pyrolysis and sol-gel [48,49]. However, the elevated cost of these processes is a barrier to the development and large scale production of SOFCs.

In the few decades past, nanostructured materials called the attention of scientific community. Earlier works suggested that the use of nanometrical particles can change the electrical properties, especially when sizes are smaller than 10 nm [50]. However, it is difficult to process them for a high densification. This limitation has hampered the obtaining and the characterization of these high density materials with small crystal size (below 20 nm). For this reason, there is not much published researches characterizing this type of material. In the last years, researches have been developed at Federal University of Rio Grande do Sul, to obtain yttria-stabilized zirconia thin films by spray pyrolysis. Depending on heat treatment, the material was obtained with crystallite size between 7 nm (when the temperature of heat treatment is 700 °C) and 70 nm (when the temperature of heat treatment is 1,200ºC) as it was observed by transmission electronic microscopy (TEM) (Figure 5.3). These studies revealed that it is possible to obtain YSZ nanoparticles when zirconium cubic phase is stabilized at minor temperature.

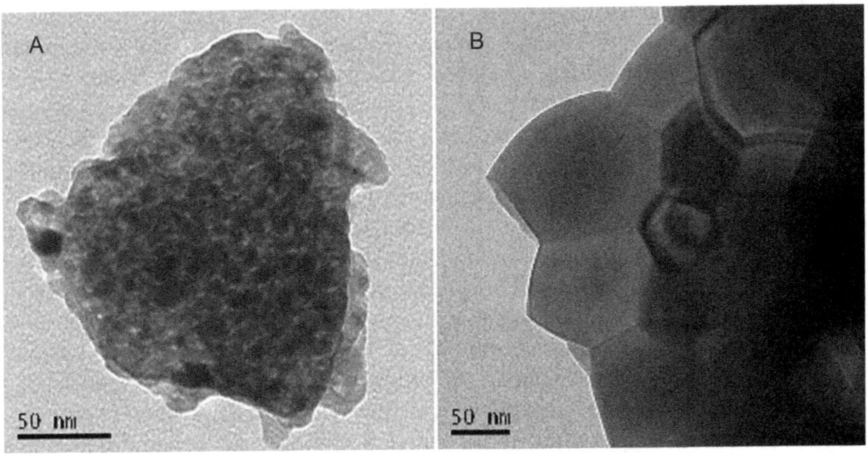

Fig. 5.3. TEM of yttria-stabilized zirconia obtained by spray pyrolysis. (a) Heat treatment at 700°C for 2h. (b) Heat treatment at 1200°C for 2h.

State of the Art Cathode

The cathode in a SOFC must transport the electrons to the reaction sites and perform the catalysis of the oxygen molecule reduction, thus, properties like electronic conductivity and catalytic activity are very important. Besides, these components must exhibit a porous (density around 70%) and homogeneous microstructure during the fuel cell operation in order to promote gas exchange within the material [44]. When choosing materials for such application it must be considered the fuel cell operation temperature, its design and the electrolyte material since it must exist a physical and chemical compatibility between the fuel cell's materials [51]. Therefore, for cells that operates at high temperatures with YSZ electrolytes, strontium-doped lanthanum manganites (LSM) are the most potential material for the cathode because of its thermal stability and high compatibility with YSZ [52]. In these cells, the reduction reaction must occur in the Triple Phase Boundary (TPB), formed by oxygen/electrolyte/cathode interface, which means that the ion created in the cathode surface is already in contact with the electrolyte and it is, hence, transported to the anode.

For IT-SOFC's LSM no longer have good properties for the application due to the poor catalytic activity of this material below 800°C. Instead, still using YSZ electrolytes, materials with ionic conductivity are most suitable and these Mixed Ionic and Electronic Conductor (MIEC's) leaded to a new class of fuel cell's cathodes, once the ion transportation can occur from the cathode to the electrolyte, creating several more reactive sites [53].

Recently, materials with layered perovskite structure are also being studied for application in SOFC, owing to its good ionic conduction. In fact, some compositions of such structure, that do not exhibit electrical conductivity, are emerging as potential electrolyte materials. Obviously, for a cathode, these materials are MIECs with more adequate ionic conduction. In despite of these good characteristics, in comparison with doped perovskites, double and layered perovskites are less stable in the operation temperatures and present a small catalytic activity which can lead to polarization losses [54].

Systems using other electrolytes besides YSZ are also being studied and thus new compatible cathodes are being developed. For cerium oxide and lanthanum gallate-based electrolytes doped barium and doped lanthanum cobaltites are promising materials for the cathode [55,56,57], however, the thermal stability of these systems has not been well evaluated, which can drastically limit the fuel cell life time. Nevertheless, these are good candidates for alternative IT-SOFC's systems.

Some good advantages emerge when a nanostructured cathode material is used. The catalysis of O_2, which is a limiting factor in the overall reaction of the fuel cell, is improved due to a considerable increase in the specific surface area which increases the reduction sites, leading to cathodes with better performance [58]. Also, as nanopowders shows better sinterability, a class of so-called composite cathodes aroused interest after some attempts to impregnate cathode material powders in electrolyte material substrates, aiming to increase the amount of TPB zones. However, at high sinterization temperatures, these materials tend to react forming undesirable phases. Hence, nanosized powder inhibits the reactions between cathode and electrolyte materials by allowing lower sinterization temperatures than regular powder [59].

Independent of the operation temperature, all cathode materials are doped or layered perovskites and, in order to obtain these materials, different methods can be used. The chosen method must promote the obtaining of a single phase perovskite, since almost all other phases formed are dielectrics, and also reach a high-specific surface area, increasing the reduction sites. In this sense, sol-gel, co-precipitation [60,61] and, combustion synthesis are the most useful methods conferring porous characteristics to the obtained powders with fine and homogenous particles. Besides, an effective doping of the perovskite [62,63] and the obtaining of nanoscale particles and nanocrystalline pure phase materials are possible through these methods [64,65]. According to the combustion synthesis parameters used, it can result in the formation of nanopores in the grain boundaries and within the particles, as demonstrates Figure 5.4, in which LSM powders were obtained using urea as fuel in 1:2 (a) and 1:4 (b) stoichometric amounts. Comparing these TEM micrographs, it is possible to observe how the over excess of fuel (1:4 ratio) in the synthesis can promote the formation of nanopores. These nanopores, when in significant amounts, not only increase the superficial surface area, but also they can create an intricate path for the air flow, promoting a more intimate contact between the cathode and the oxygen. Besides this, it is known that these open nanopores can inhibit grain growth as well as pores prevent grain coarsening in ceramics with ordinary grain size (>100 nm) [66].

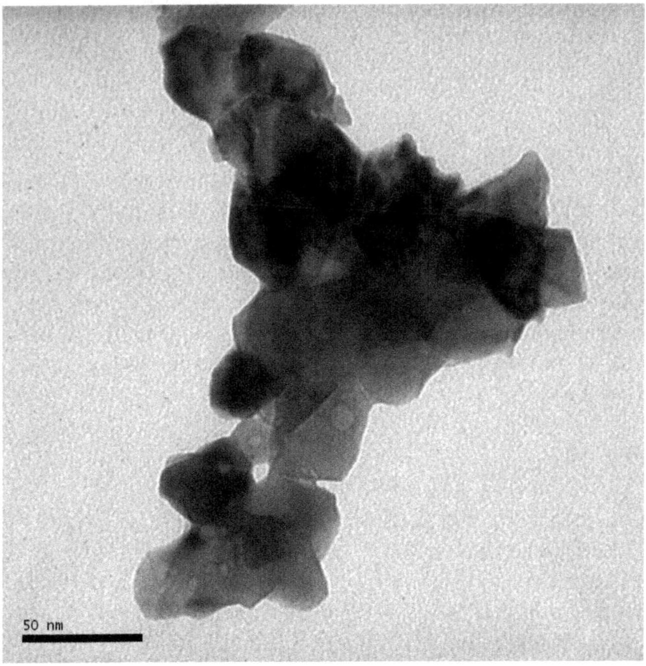

(a)

Fig. 5.4. TEM micrographs of LSM powders obtained by urea combustion synthesis in stoichometric fuel ratios of 1:2 (a) and 1:4 (b), illustrating the influence the fuel amount in the formation of nanopores.

(b)

Fig. 5.4. (*Continued*)

In the SOFC's 3rd generation, with focus in both economical and technical issues, the fuel cell components are thin films formed by processes that often start with solid in suspension. As said, the cathode porosity must be around 30%, diverging from a very usual problem that comes along when the formation of a nanosuspension is needed for processing these materials, which is the difficult achievement of high density depositions [67]. Hence, it is likely that a more accurate handling of the suspension is possible, enabling the use of greater amounts of solids in the solution.

Art state of Anode

The anode is an important part of the heart cell since it is through it that the fuel (hydrogen, methanol, ethanol, and others) will be provided. The anode must have at least two requirements: (i) must have high catalytic activity in order to obtain high current densities and (ii) must have controlled porosity. Nowadays, the scientific community wants to build a cell that is compatible with the use of liquid fuels such as ethanol, because it is easy handling and transport. However, the main problem that hinders the oxidation of ethanol in the SOFC, is the fact that their presence at high temperatures can lead to carbon deposition on the surface of the anode. Ni-based catalysts are suitable for the cracking reaction of hydrocarbons, but are very susceptible to coating with carbon, resulting in rapid and irreversible degradation of the cell [40]. Another problem in this type of material has been the low compatibility of nickel with the electrolyte. According to Priyatham [68], this issue can be solved by a mixture with anode materials and electrolyte materials. Besides these problems, the SOFC anode side undergoes a sintering due to high temperature, causing a closure of pores and irreversible decline the efficiency of SOFC. To minimize it, several studies have been performed with the addition of chromium in the cermet used as anode. This addition has given a very positive outcome, besides the performance of the cell depended more of the performance of the cathode than the anode [41].

5.3 Concluding Remarks

Tendency for medium and long term is increase the use of alternative and sustainable energy sources, saving the world of environmental catastrophe. Researches on nanostructured materials broads the development of new ways to capture, store, and transfer energy with high performance and low cost. It may be useful to view this specific discipline more as a process-based technology that is redefining the entire product landscape enabling smaller and more efficient devices. Inorganic nanoparticles and nanocrystals, nanowire technology, carbon nanotube technology, and nanoporous materials based on template technology are few examples of the last advancements for highly efficient batteries, flexible solar cells and portable fuel cells with low cost, for example.

Improving the efficiency of fuel cells through the use of nanotechnology appears to be more plausible by using molecularly tailored catalysts, polymer membranes, and improved fuel storage. In a long-term range, SOFC's technology may become the best alternative to obtain energy without damage to the environmental.

References

[1] http://www.a123systems.com/ Accessed June 2010
[2] http://www.nanotecture.co.uk/home.aspx Accessed June 2010
[3] http://www.toyota.com/about/environment/innovation/advanced_vehicle_technology/ Accessed June 2010
[4] http://www.valence.com/ Accessed June 2010
[5] http://www.altairnano.com/profiles/investor/fullpage.asp?f=1&BzID=546&to=cp&Nav=0&LangID=1&s=0&ID=10724 Accessed June 2010
[6] http://www.next-alternative.com/technology.html Accessed May 2010
[7] Kohzaki M, Carbon-containing lithium-iron composite phosphorus oxide for lithium secondary battery positive electrode active material and process for producing the same. US 7025907, 2006
[8] Zhang P, High performance anode material for lithium-ion battery. US 7722991, 2010
[9] Kiebele A, Gruner G (2007) Carbon nanotube based battery architecture. Appl Phys Lett 91:144104.1-144104.3

[10] Park M-H, Kim MG, Joo J, Kim K, Kim J, Ahn S, Cui Y and Cho J (2009) Silicon Nanotube Battery Anodes. Nano Letters 9: 3844-3847
[11] http://www.maxell.co.jp/e/products/rd/catalyst.html Accessed May 2010
[12] http://www.nanoener.com/app.html Accessed June 2010
[13] http://www.cyriumtechnologies.com/ Accessed May 2010
[14] http://www.illuminex.biz/ Accessed June 2010
[15] http://www.nanosolar.com/ Accessed June 2010
[16] http://www.unidym.com/products Accessed May 2010
[17] http://www.ceramatec.com/technology/ Accessed May 2010
[18] http://www.cleantechnano.com/applications.html Accessed May 2010
[19] http://www.nextechmaterials.com/ Accessed May 2010
[20] http://www.usnanocorp.com/ Accessed May 2010
[21] Kang H-R, Electrode for fuel cell, method of producing the same, and fuel cell including the electrode. US 7718304 B2, 2006
[22] http://nanopatentsandinnovations.blogspot.com/2010/03/samsung-fuel-cell-with-rotating-disk.html Accessed May 2010
[23] http://www.mtimicrofuelcells.com Accessed May 2010
[24] http://www.polyfuel.com Accessed May 2010
[25] http://www.smartfuelcell.de Accessed June 2010
[26] Stambouli A B, Traversa E (2002) Solid Oxide Fuel Cells (SOFCs): A Review of an Environmentally Clean and Efficient Source of Energy. Renew Sust Energ Rev 6: 433-455
[27] http://www.fuelcellenergy.com/files/FCE%20WhitePaper%20040308_2.pdf Accessed May 2010
[28] Kirubakaran A, Jain S, Nema RK (2009) A review on fuel cell technologies and power electronic interface. Renew Sust Energ Rev 13:2430-2440
[29] Hoogers G (2002) Introduction. In: Hoogers G (ed) Fuel Cell Technology Handbook, 1st edn, Taylor & Francis
[30] Steele BCH (2000) Materials for IT-SOFC stacks 35 years R&D: the inevitability of gradualness? Solid State Ionics 134:3-20
[31] Tucker MC (2010) Progress in metal-supported solid oxide fuel cells: A review. J of Power Sources 195: 4570-4582
[32] Perednis D, Gauckler LJ (2004) Solid oxide fuel cells with electrolytes prepared via spray pyrolysis. Solid State Ionics 166:229-239
[33] Hagiwara A, Hobara N, Takizawa K, Sato K, Abe H, Naito M (2007) Microstructure control of SOFC cathodes using the self-organizing behavior of LSM/ScSZ composite powder material prepared by spray pyrolysis. Solid State Ionics 178: 1123-1134
[34] Mauvy F, Bassat J-M., Boehm E, Manaud J-P, Dordor , Grenier J-C (2003) Oxygen electrode reaction on $Nd_2NiO_{4+\delta}$ cathode materials: impedance spectroscopy study. Solid State Ionics 158:17-28
[35] Kosacki I, Rouleau CM, Becher PF, Bentley J, Lowndes DH (2005) Nanoscale effects on the ionic conductivity in highly textured YSZ thon films. Solid State Ionics 176:1319-1326
[36] Antoni L (2004) Materials for solid oxide fuel cells: the challenge of their stability. Mater Sci Forum 461-464:1073-1090

[37] Fontana S, Amendola R, Chevalier S, Piccardo P, Caboche G, Viviani M, Molins R, Sennour M (2007) Metallic interconnects for SOFC: Characterisation of corrosion resistance and conductivity evaluation at operating temperature of differently coated alloys. J of Power Sources 171:652-662
[38] Fergus JW (2005) Metallic interconnects for solid oxide fuel cells. Mater Sci Eng A 397:271–283
[39] http://revistapesquisa.fapesp.br/index.php?art=3141&bd=1&pg=1&lg= Accessed May 2010
[40] Ivers-Tiffee E, Weber A, Herbstritt D (2001) Materials and technologies for SOFC-components. J Eur Ceram Soc 21:1805-1811
[41] Jiang SP, Chan SH (2004) A review of anode materials development in solid oxide fuel cells. J Mater Sci 39: 4405-4439
[42] Huijsmans JPP (2001) Ceramic in Solid Fuel Cells. Curr Opin Solid State Mater Sci 5:317-323
[43] Boulc'h F, Djurado E (2003) Structural changes of rare-earth-doped, nanostructured zircônia solid solution. Solid State Ionics 157:335-340
[44] Florio DZ, Fonseca FC, Muccillo ENS et al (2004) Materiais Cerâmicos para Células a Combustível. Cerâmica 50:275-290
[45] Sasaki K, Muranaka M, Suzuki A, Terai T (2008) Synthesis and characterization of LSGM thin film electrolyte by RF magnetron sputtering for LT-SOFCS. Solid State Ionics 179:1268-1272
[46] He Z, Yuan H, Glasscock JA, Chatzichristodoulou C, Phair JW, Kaiser A, Ramousse S (2010) Densification and grain growth during early-stage sintering of $Ce_{0.9}Gd_{0.1}O_{1.95-\delta}$ in reducing atmosphere. Acta Mater 58:3860-3866
[47] Gaudon M, Laberty-Robert Ch, Ansart F, Stevens P (2006) Thick YSZ films prepared via modifited sol-gel route: Thickness control (8-80μm). J Eur Ceram Soc 26: 3153-3160
[48] Perednis D (2003) Thin Film Deposition by Spray Pyrolysis and the Application in Solide Oxide Fuel Cells. Thesis - Suiss Federal Institute of Technology Zurich
[49] Todorovska R, Petrova N, Todorovsky D (2005) Spray pyrolysis deposition of YSZ and YSZ-Pt composite films. Appl Surf Sci 252:1266-1275
[50] Chiodelli G, Maglia F, Anselmi-Tamburini U et al (2009) Characterization of low temperature protonic conductivity in bulk nanocrystalline fully stabilized zirconia. Solid State Ionics 180:297-301
[51] EG&G Technical Services (2002) Fuel cell handbook. US Department of energy, West Virginia
[52] Sahu A K, Ghosh A, Suri A K (2009) Characterization of porous lanthanum strontium manganite (LSM) and development of yttria stabilized zirconia (YSZ) coating. Ceram Int 35:2493–2497
[53] Richter J, Holtappels P, Graule T et al. (2009) Materials design for perovskite SOFC cathodes. Monatsh Chem 140:985–999
[54] Tarancon A, Burriel M, Santiso J et al (2010) Advances in layered oxide cathodes for intermediate temperature solid oxide fuel cells. J Mater Chem 20:799–3813
[55] Jena H, Rambabu B (2007) An exploratory study on solution assisted synthetic routes to prepare nano-crystalline $La_{1-x}M_xGa_{1-y}N_yO_{3\pm\delta}$ (M=Sr□; N= Mn, Mg) for IT-SOFC applications. Mater Chem Phys 101:20–29
[56] Liu B, Tanga L, Zhanga Y (2008) Preparation and characterization of $La_{0.9}Sr_{0.1}Ga_{0.8}Mg_{0.2}O_{3-\delta}$ thin film on the porous cathode for SOFC. Int J Hydrogen Energ 34:440-445

[57] Zhu C, Liu X, Yi C et al (2009) Novel $BaCo_{0.7}Fe_{0.3-\gamma}Nb_yO_{3-\delta}$ (y = 0–0.12) as a cathode for intermediate temperature solid oxide fuel cell. Electrochem Commun 11:958-961

[58] Darbandi AJ, Hahn H (2009) Nanoparticulate cathode thin films with high electrochemical activity for low temperature SOFC applications. Solid State Ionics 180:1379–1387

[59] Chen J, Liang FL, Liu LN, Jiang SP, Chi B, Pu J, Li J (2008) Nano-structured (La, Sr)(Co, Fe)O_3 + YSZ composite cathodes for intermediate temperature solid oxide fuel cells. J Power Sources 183:586–589

[60] Gaudon M, Laberty-Robert C, Ansart F, Stevens P, Rousset A (2002) Preparation and characterization of $La_{1-x}Sr_xMnO_{3+\delta}$ ($0 \leq x \leq 0.6$) powder by sol–gel processing. Solid State Sci 4:125–133

[61] Ghosh A, Sahu AK, Gulnar AK, Suri AK (2005) Synthesis and characterization of lanthanum strontium manganite. Scripta Mater 52:1305–1309

[62] Manoharan SS, Patil KC (1993) Combustion Route to Fine Particles Perovskite Oxides. J Solid State Chem 102:267-276

[63] Prabhakaran K, Joseph J, Gokhale NM, Sharma SC, Lal R (2005) Sucrose combustion synthesis of $La_\chi Sr_{(1-\chi)}MnO_3$ ($\chi \leq 0.2$) powders. Ceram Int 31:327–331

[64] Tyagi AK, Chavan SV, Purohit RD (2006) Visit to the fascinating world of nanoceramic powders via solution- combustion. Indian J Pure Ap Phy 44:113-118

[65] Deganello F, Marcì G, Deganello G (2009) Citrate–nitrate auto-combustion synthesis of perovskite-type nanopowders: A systematic approach. J Eur Ceram Soc 29: 439-450

[66] Groza JR (1999) Nanosintering. Nanostruct Mater 12:987-992

[67] http://www.azom.com/details.asp?ArticleID=919#_Design_Problems Accessed August 2010

[68] Priyatham T, Ranjit R (2010) Synthesis and characterization of nanocrystalline Ni-YSZ cermet anode for SOFC. Mater Charact 61:54-58

Abbreviations

CNTs - carbon nanotubes
PHEVs - Plug-In Hybrid Electric Vehicles
DMFC - direct methanol fuel cell
PEFC - polymer electrolyte fuel cell
QDEC - Quantum Dot Enhanced Cell
CNSC - Carbon NanoSphere Chains
FCHV - Fuel Cell Hybrid Vehicle
LSGM - Lanthanum Strontium Gallate Magnesite
SYT - yttrium doped strontium titanate
MEA- membrane electrode assembly
SOFCs - solid oxide fuel cells
TPB - triple phases boundaries
ID - internal diffusion
ITSOFC - Intermediate Temperature Solid Oxide Fuel Cell
YSZ - yttria-stabilized zirconia
LSGM - Strontium and magnesium doped lanthanum galatte
CGO - gadolinium-doped ceria
PVD - physical vapor deposition
CVD - chemical vapor depositions
TEM - transmission electronic microscopy
LSM - strontium-doped lanthanum manganites
MIEC's - Mixed Ionic and Electronic Conductor
AFC - alkaline fuel cell
MCFC - molten carbonate fuel cell
PAFC - phosphoric acid fuel cell
PEMFC - proton exchange membrane fuel cell
SOFC - solid oxide fuel cell

6 Materials for Bio-Applications

Alice Gonçalves Osorio, Michelle Dunin-Zupanski, and Rafael Mello Trommer

Departamento de Engenharia de Materiais, Universidade Federal do Rio Grande do Sul, 90035190, Porto Alegre, Brazil
E-mail: michelledunin@yahoo.com.br

Abstract. The recent advances in nanomaterials have brought potential applications in several fields of medicine and health care, such as diagnosis, treatment of diseases, tissue engineering, etc. This chapter introduces the main nanomaterials used in bio-applications as well as potential materials for future applications in biomedicine. Magnetic nanoparticles such as iron oxide are considered as one of the most promising nanomaterials. Hydroxyapatite is the main bioceramic used in orthopedics and dentistry, due to its chemical similarity to the inorganic fraction of the human bone. Carbon nanotubes present great potential for bio-applications such as biomaterials, drug delivery and tissue engineering. Metals like silver nanoparticles are potential antimicrobial agent, mainly because of their large surface area. Nanomaterials are being widely used and studied in bio-applications because of their novel and unique physicochemical properties, compared to bulk materials. Their unique properties, however, can present an unpredictable impact on human health. Therefore, the toxicology of nanomaterials has attracted much attention worldwide.

Keywords: Apatite, Nanohydroxyapatite, Wet chemical precipitation, Sol–gel, Biomimetic deposition.NanoHydroxyapatite, Zinc oxide, Magnetic nanoparticles (MNPs), Organic modified silica (Ormosil), Silver and gold nanoparticles, Carbon nanotubes, Toxicity, Drug delivery.

Cells are the main constituent of the living organisms, with a typical size of 10 μm. They can be, however, much smaller even in the sub-micron size domain. With a typical size of 5 nm, proteins are smaller than cells, with dimensions comparable to the smallest nanoparticles that have been made by researchers. This circumstance has led to the improvement of the health care and medical research through the development of nanoproducts, as a result of private and public research efforts.

Nanomaterials have been used in bio-applications such as cosmetics, drug delivery, imaging and medical diagnosis, tissue engineering, etc. However, only few products have entered the marketplace. Some companies, e.g., have been working on the development of nanomaterials for bio-applications in the pharmaceutical field, mainly for drug delivery. Other companies are applying ceramic materials in the nanometric size to tissue engineering and orthopedics.

Up-to-date nanomaterials are being used to improve the effectiveness of medical diagnosis and therapeutics, such field was named nanomedicine. The potential use of nanomaterials in bio-applications is enormous. However, there are still several questions about their long-term safety and the risk–benefit characteristics of their usage.

6.1 Ceramics

The class of ceramics used for bio-applications is referred to as bioceramics. Bioceramics are mostly used as orthopedic implants and in dental applications to repair and replace diseased and damaged parts of the musculoskeletal system. Nanostructured bioceramics mainly studied and commercialized to date are hydroxyapatite, zinc oxide and organic modified silica; these materials are therefore detailed below [1].

6.1.1 Hydroxyapatite

Hydroxyapatite (HA) is the main inorganic constituent of bones and teeth [2], [3]: the inorganic fraction of the human bone consists of an amorphous phase of tricalcium phosphate and a crystalline phase of HA with nanometer-sized and needle-shaped crystals of approximately 5 - 20 nm [4]. The enamel of human teeth contains about 95% of hydroxyapatite [5].

Due to the chemical similarity between HA and mineralized human bone, synthetic HA exhibits strong affinity to host hard tissues. Also due to its chemical nature, HA is capable to adsorb molecules of water and proteins, allowing strong chemical bonding between HA and surrounding bone tissue under *in vivo* conditions [6].

Nanotechnology has opened up innovative techniques for producing bone-like synthetic nanopowders and hydroxyapatite coatings. New developments have brought new opportunities to design superior biocompatible coatings for implants, and the development of high-strength dental and orthopedic nanocomposites for medical applications [4].

Nanohydroxyapatite Characteristics

Natural apatite minerals are found in igneous, sedimentary and metamorphic rocks, while biological nano-sized apatites are found in bones and teeth of vertebrates [3]. Biological apatites, known as calcium hydroxyapatite, differ from pure HA in characteristics such as stoichiometry, composition and mechanical properties. Pure hydroxyapatite is not found in biological environment. Biological hydroxyapatites are described as calcium-deficient and non-stoichiometric, as a result of various substitutions at regular HA points [7].

Hydroxyapatite is a compound with chemical formula of $Ca_{10}(PO_4)_6(OH)_2$ and a definite ratio Ca/P of 1.67. The term apatite describes a family of minerals with general formula of $M_{10}Y_6X_2$, where M represents metals such as Na, K and Mg; Y represents functional groups such as PO_4, CO_3 and sulfates; and the X stands for atoms or groups such as OH, F, and Cl. Therefore, apatite is a structural description, and not a composition [7], [8].

Hydroxyapatite belongs to the hexagonal system, with space group $P6_3/m$ (characterized by a six fold c-axis perpendicular to three equivalent a-axes at angles 120° to each other [8]), with cell dimensions of a = 9.423 and c = 6.875 (valid for mineral or pure HA sintered at 1100°C). There are six PO_4 groups and two OH groups, as well as two chrystallographically independent Ca atoms in the HA unit cell [3].

Nanohydroxyapatite (Fig 6.1) has extremely high surface area. Because the atoms in the surface layer have unsaturated atomic bond, nano-HA exhibit a high bioactivity, which accelerates the early stage bone growth and tissue healing.

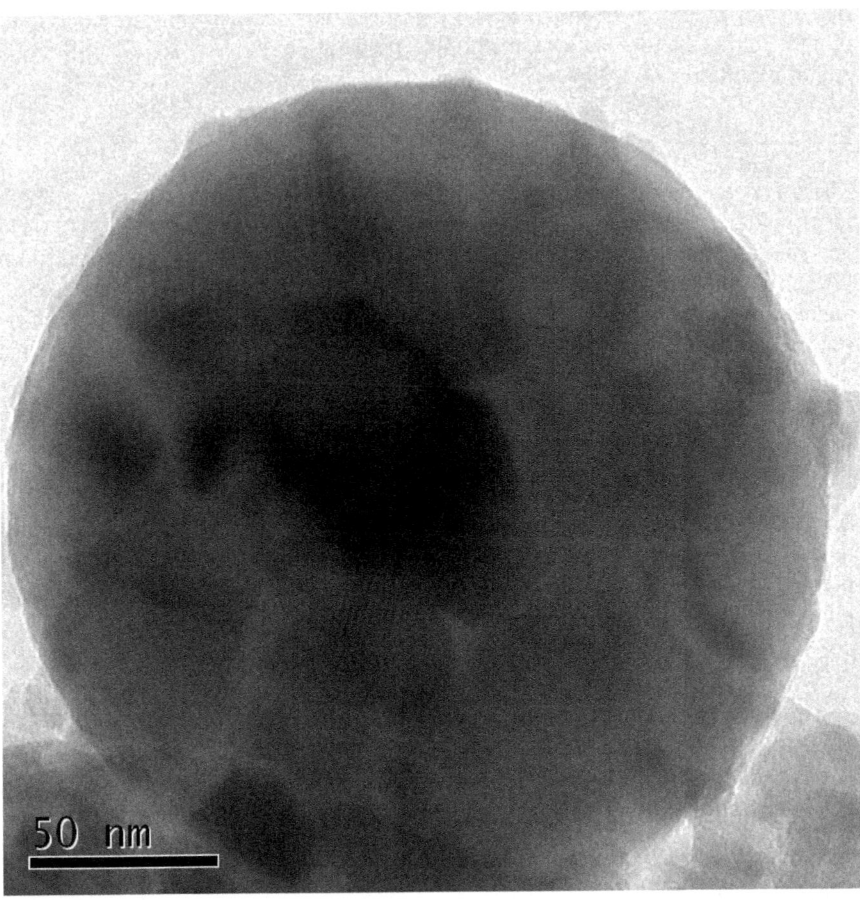

Fig. 6.1. Scanning Electron Microscopy (SEM) image of nanostructured spherical HA powder.

According to calculations made by Zhang and co-workers [9], for 10 nm grain size, the surface atoms account for about 22%, while 50 nm grains have only 5% atoms on the surface. If the grain size is larger than 100 nm, the surface atoms account for less than 2.5%. The smaller the grain size, the higher is the number of surface atoms, resulting in quicker bone growth and faster dissolution rate. Therefore, nano-HA is expected to induce different biological responses, due to higher solubility when compared to micrometric synthetic HA.

Solubility is probably the most important property of calcium phosphates for *in vivo* evaluation, since the *in vivo* behavior can be predicted to a large extent by its solubility in water. If the solubility of the compound is lower than the mineral part of the bone, degradation is extremely low. Under normal physiological conditions (pH = 7.2), hydroxyapatite is the stable calcium phosphate [10]. Hydroxyapatite is soluble in acidic solutions and poorly soluble in water, while insoluble in alkaline solutions. However, the solubility of hydroxyapatite in water increases with the addition of electrolytes. Furthermore, the solubility of hydroxyapatite is modified in the presence of amino acids, proteins, enzymes and other organic compounds. Other factors such as method of preparation and density also affect the solubility of HA [3].

Preparation Methods

There are various methodologies employed to produce nanostructured synthetic HA, including wet chemical synthesis (such as wet precipitation [11], [12], [13], and sol–gel techniques [14], [15], [16], [17], [18], [19]) and biomimetic deposition [20]. It is important to note that each method is greatly dependent on the purity and quality of reactants. Bernard *et al.* [21] investigated the effects of purity of the precursors on the final HA product. They established that the purity of reactants, in addition to proper stoichiometry and subsequent thermal treatment have direct influence on the quality of the final product.

Synthesis via wet chemical precipitation

An HA nanocrystal suspension can be prepared by a wet chemical precipitation reaction [6]. There are two most popular methods of precipitating HA. One is a process involving an alkaline solution of Ca precursor and an acid solution of P precursor. The chemical formula for this process is as follows:

$$10Ca(OH)_2 + 6H_3PO_4 \rightarrow Ca_{10}(PO_4)_6(OH)_2 + 18H_2O \quad (6.1)$$

The other method was proposed by Hayek and Newesley in 1963 [22], and involves calcium and phosphate salts. Ammonium hydroxide is added to ensure a highly alkaline pH to the reaction, which induces the formation of stoichiometric HA after sintering [8]. The Hayek and Newesley chemical formula is:

$$10Ca(NO_3)_2 + 6(NH_4)_2HPO_4 + 8NH_4OH \rightarrow Ca_{10}(PO_4)_6(OH)_2 + 20NH_4NO_3 + 6H_2O \quad (6.2)$$

The shape, size and specific surface area of the HA nanoparticles obtained by wet precipitation are very sensitive to the reactant addition rate, and to the reaction temperature. Also, parameters of calcinations and ripening, such as, time and temperature all have significant influence on the final quality of the HA, as demonstrated by the studies of Pang and Bao [23].

Synthesis via sol–gel

The sol–gel process allows low-temperature formation of nanocrystalline HA. While temperatures of about 1000°C are usually required to sinter apatite crystals prepared from wet precipitation, the sol–gel route allows HA formation with temperature as low as 600°C [14], [15]. The method also offers a molecular-level mixing of precursors, improving the chemical homogeneity of the resulting HA to a significant extent [18]. In addition, the sol–gel technique offers an option for homogeneous HA coating on metal substrates such as stainless steel [24] and titanium [25].

There are a number of Ca and P precursor combinations employed for sol–gel synthesis. However, it is important to note that a major limitation in the sol–gel technique is associated with the hydrolysis of the phosphorous precursors. Phosphorus alkoxides are frequently used as the phosphorus precursors for sol–gel HA synthesis, such as triethyl phosphate and triethyl phosphite. The hydrolysis activity of these precursors is relatively poor and a prolonged time period (of several days for triethyl phosphate) is needed to form the HA phase [26].

Biomimetic deposition

Metastable synthetic body fluids (SBF), with an inorganic salt composition similar to that of human blood plasma, at physiological pH and temperature, allows the spontaneous nucleation and growth of a nanosized carbonated HA [20]. The process is very simple and low cost, and was first described by Kokubo [27]. In his studies, Kokubo soaked a glass-ceramic sample in a SBF solution, coating the surface of the sample with a thin layer of "bone-like" carbonated apatite. Since Kokubo's reports, this method has been proven to facilitate spontaneous generation and growth of biomimetic HA on immersed silica-gels, bioglass and titanium samples [6].

Commercial Nanohydroxyapatite

HA has been used for a variety of biomedical applications since the 1970s, including artificial bone grafts, coating for metallic prosthesis and matrices for drug release control. The major products used are coatings on metallic dental, hip, and spine implants for the acceleration of early stage healing and decreasing the pain. Other products such as nano-HA powders and porous blocks are used as bone fillers [28]. Due to the chemical similarity between HA and mineralized bone the synthesized nano-HA is expected to be recognized as belonging to the body. In other words, the synthesized nano-HA could be directly involved in the natural bone remodeling process. Formation of chemical bond with the host tissue offers HA a greater advantage in clinical applications over most other bone substitutes.

A number of companies have marketed hydroxyapatite for medical use since 1980s, but only recently nanometer HA has been introduced in the market. Some nano-HA commercially available are listed on Table 6.1.

Table 6.1. Commercial Hydroxyapatites

Company	Product name	Specific Surface Area (m^2/g)	Average Particle Size (nm)
Berkeley Advanced Biomaterials [29]	BABI-HAP-N20-A	110	20 – 70
	BABI-HAP-N100	50 – 70	100
	BABI-HAP-N550-D	10 – 50	550
Fluidinova [30]	nanoXIM – Hap100	> 100	< 50
Sigma – Aldrich [31]	702153 Hydroxyapatite	14.3	< 200
	677418 Hydroxyapatite	> 9.4	< 200
	693863 Hydroxyapatite	10 – 15	< 200

Hydroxyapatite is not yet commercialized as drug carrier. These systems are still being improved for future release in the biomedical market. However, patents describing HA as carrier for drug substances are available, such as US Patent 6558703 [32] and US Patent 6767550 [33].

6.1.2 Zinc Oxide

Ultraviolet inorganic filters in sunblocks

One of the main uses of zinc oxide (ZnO) as material with bio-application is as ultraviolet inorganic filters in sunblocks. The increase in the popularity of zinc oxide as inorganic filter is because of its insoluble particles that remain on the skin and is not absorbed systematically [34] . In this case, the light is either absorbed into the sunblock material or reflected away from the skin, similar to a mirror. Zinc oxide provides physical protection from damaging rays, absorbing primarily ultraviolet A (UVA) light rather than scattering or reflecting it.

A sunblock that uses mineral inorganic particles as zinc oxide, nonetheless, presents the tendency to reflect the light and, therefore, turn the product into an opaque and white sunblock, which makes its use as a cosmetic impossible [35]. One way to avoid this undesired effect is by the use of a nanosized ZnO, with a particle size between 20 and 50 nm, thus avoiding the opacity and whitening of the sunblock [36] .

Several methods have been used to produce nanosized ZnO, such as thermal decomposition, flame spray pyrolysis and precipitation method. Hydrothermal synthesis is one of the main processes used to produce zinc oxide nanoparticles, it is used by companies such as Nanostructured & Amorphous Materials [37], SB Chemical Co. Ltd.

[38], NanophaseTechnologies Corp (NanoArc™) [39], Micronisers Pty Ltd (Microsun™) [40] and Evonik Industries [41]. An alternative and cost-effective technique to produce nanoparticles of zinc oxide comprises the use of a flame where an atomized precursor solution is sprayed. After the chemical reactions, the ZnO powder is obtained with a particle size close to 33 nm, as shows Fig 6.2 [42].

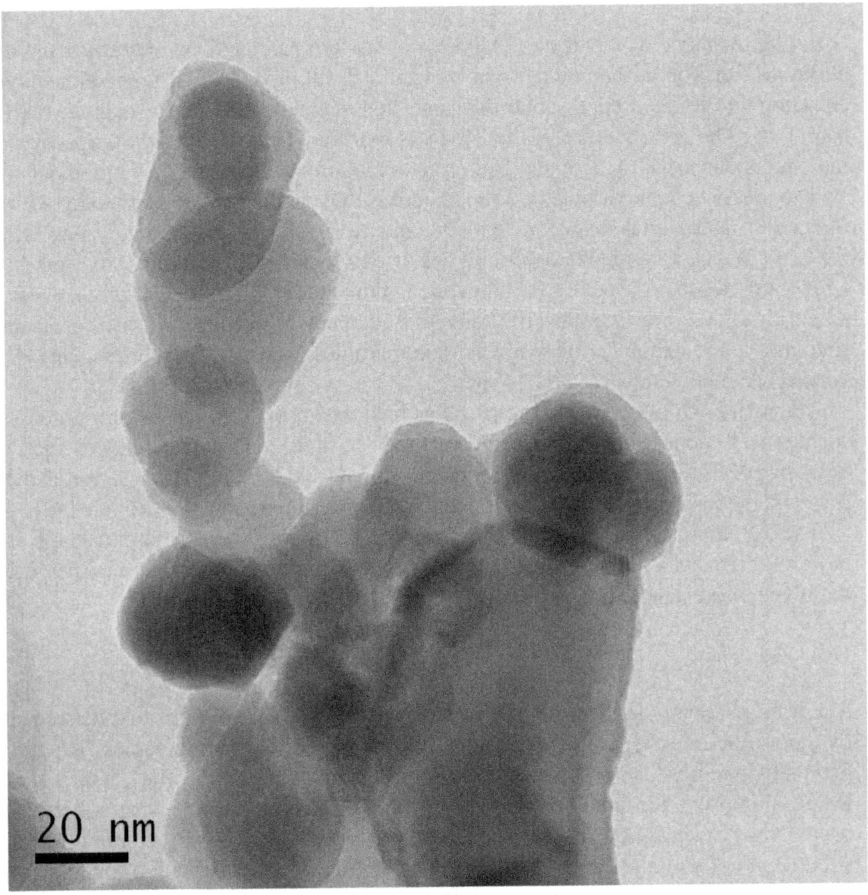

Fig. 6.2. Nanoparticles of zinc oxide prepared by the alternative method reported by Trommer *et al.* [42].

Antibacterial treatment

Another important use of zinc oxide in bio-applications is as antibacterial agent. The advantages of ZnO are the small amount required for the antibacterial activity as well as the pH values between 7 and 8 that allows it to be used as water for washing and drinking [43]. An example of the potential use of ZnO as a nanomaterial in industry is the coating of fabrics. In this case, the antibacterial activity of ZnO can be increased with the decrease of the particle size [44].

6.1.3 Magnetic Nanoparticles (Fe_3O_4 and γ-Fe_2O_3)

Magnetic nanoparticles have been successfully used in medicine for pathogen detection, proteins separation, drug delivery and hyperthermia [45].

Imaging

Once the magnetic nanoparticles (MNPs) possess unique magnetic properties and the ability to function at the cellular and molecular level of biological interactions, they are attractive to be used as contrast agents for magnetic resonance imaging (MRI) [46], [47]. The great feature of the MNPs is that they become a single magnetic domain independent of the size and, therefore, maintain one large magnetic moment.

The magnetic nanoparticles are the materials that can be manipulated under the influence of an external magnetic field. Magnetite Fe_3O_4 and maghemite γ-Fe_2O_3 are the main magnetic oxide nanoparticles because of their biocompatibility and stability, which justifies their large use on biomedical applications. Commercial products such as Feridex I.V.^{TM}, a liver-specific magnetic resonance imaging contrast agent, and Endorem^{TM} are examples of magnetic nanoparticles produced by the company Advanced Magnetic Industries, Inc [48].

Magnetite and maghemite nanoparticles have been produced by several processes such as wet chemistry solution-based methods to more complex techniques such as laser pyrolysis or chemical vapor deposition. Iron-oxide nanoparticles can also be prepared by co-precipitation of Fe(III) and Fe(II) with ammonium hydroxide [49].

The wet chemical routes to magnetic nanoparticles are simpler, more tractable and more efficient with appreciable control over size, composition and sometimes even the shape of the nanoparticles [50].

Drug delivery

Magnetic nanoparticles may also be used to activate or release drugs to a desired site by the use of an external magnetic field, as e.g., in novel cancer therapies. The magnetic nanoparticles can provide a mechanism for the guidance of drugs, which is believed to improve the safety and efficacy of the drug and reduce the side effects. The drug delivery of magnetic nanoparticles comprises the guidance of the drug to the desired locale of action, where a high dose of the drug is provided with a low systemic concentration. Currently, the company Biophan [51] is developing nanomagnetic carriers designed to attach to and precisely deliver a drug to a target site and activate the drug only in that area. This can provide a high local dose with low systemic concentration and enable improved safety profiles, reduced toxic side effects, and deliver clinical doses higher than what is currently possible.

6.1.4 Organic Modified Silica (Ormosil)

The photodynamic cancer therapy destroys the cancer cells by laser-generated atomic oxygen, which is cytotoxic. However, during the treatment the remaining dye molecules can migrate to the skin and eyes, which makes the patient very sensitive to the daylight exposure. An alternative to avoid this side effect is the use o porous nanoparticle (~ 30 nm) where the hydrophobic version of the dye molecule is enclosed inside this nanoparticle [52]. One way to achieve such condition is the use of silica gels chemically modified by organic precursors, named Ormosil [53], [54]. This organically modified silica keeps the dye trapped inside the material and prevents the spread of the dye to other parts of the body while the oxygen is allowed to diffuse through the small pore size of about 1 nm [55].

6.2 Metals

Metallic biomaterials are used for load bearing applications and present sufficient fatigue strength to endure the rigors of daily activity such as walking, chewing, etc. Metallic materials are normally used as pins and plates and femoral stems, orthopedic implants, etc. Currently, the only nanostructured metals seen in the marketplace are nanoparticles.

The enhanced resonance absorption and scattering properties as well as strong Raman scattering of metallic nanoparticles such as silver, gold and copper, make them especially useful for applications such as photothermal therapy, optical imaging and Raman probe [45] .

Silver nanoparticles have been considered as a potential antimicrobial agent, mainly because of their large surface area, which allows a better contact with the microorganism [56]. The bacterial effect of silver is size dependent due to the electronic effects that enhance the reactivity of nanoparticles that will interact with the bacteria.

Due to their surface Plasmon resonance (SPR), which enhances light scattering and absorption, gold nanoparticles have enormous potential for diagnosis and therapy of cancer [57]. Moreover, gold nanoparticles provide non-toxic carriers for drug and gene delivery applications [58].

Gold nanoparticles are usually produced by the reduction method, such as the reduction of $AuCl(PPh_3)$ with diborane or sodium borohydride, biphasic reduction of $HAuCl_4$ by sodium borohydride in the presence of thiol capping agents or by the reduction of $HAuCl_4$ with sodium citrate in water.

6.3 Composites

Biomedical composites are used as an alternative to obtain specific properties through the combination of two or more materials. To date, they are mainly used for drug delivery, tissue engineering applications and cosmetic orthodontics. They tend to mimic the structures of the living materials involved in the process in addition to the strengthening properties of the matrix, but still providing biocompatibility, e.g., scaffolds in bone tissue engineering. Up to now there has little research on nanostructured biomedical composites, carbon nanotubes is on the top of the list of researches on such field.

6.3.1 Carbon Nanotubes

Carbon nanotubes (CNTs) have recently emerged in the field of biomedical materials and devices, it consists of one or more cylindrical layers of graphene with a diameter in the order of nanometers. Some of the applications that CNTs present great potential are novel biomaterials, drug delivery and tissue engineering.

The usage of CNTs in medicine can be mainly related to three of their properties: (i) mechanical resistance, where CNTs are used as reinforcement or structural support; (ii) chemical property, due to their ability to be funcionalized; and (iii) electrical conductivity.

For applications such as novel biomaterials and scaffolds the main role played by CNTs is related to their unique mechanical properties. Nanotubes are believed to be very promising in the enhancement of ceramic and polymeric matrices. They implement toughness to brittle bioceramics as well as mechanical strength to biopolymers. Synthetic hydroxyapatite (HA), e.g., is a biocompatible ceramic material usually used to coat hip prosthesis. Its poor resistance and toughness added to poor inter-laminar shear strength between coating layer and implant surface led researchers to use CNTs as reinforcement to HA. Chen and co-workers [58], e.g., deposited CNTs reinforced-HA film on Ti-6Al-4V substrate (material used to manufacture hip prosthesis) by laser surface alloying and obtained a remarkable increase in the hardness and slightly increase in the elastic modulus of the coating.

Kaya [59] also evaluated CNTs-reinforced HA film deposited by means of electrophoretic deposition. The researcher concluded, therefore, that hardness, elastic modulus and inter-laminar shear strength increased when CNTs are added to the ceramic matrix.

Other researchers such as Balani [60] and Hahn and co-workers [62] also studied the performance of this coating and obtained similar results. From the researches and results published to date, one can conclude that CNTs-reinforced HA coating is a promising material for high load-bearing metal implants.

In addition, some researchers envisaged the manufacture of a product completely made of HA reinforced with CNTs to be used as biomaterial [62], [64]; they obtained effective dispersed CNTs in HA and concluded that bending strength and fracture toughness are improved considerably.

Moreover, Xu and co-workers [65] proved the efficiency of CNTs reinforcing HA to biomedical applications by the evaluation of the population of osteoblast cells on the pure HA and CNTs-reinforced HA. Researchers showed that the presence of CNTs in HA promoted growth and adhesion of osteoblast cells.

Another field of applications of CNTs is into scaffolds as an attempt to provide structural reinforcement as well as impart novel properties such as electrical conductivity into the scaffolds. Harrison and Atala [66] reviewed the work done on CNTs reinforcing scaffolds and documented that beside structural reinforcement, CNTs may impart, accelerate and direct cell growth throughout their electrical conductivity. On top of that, for such application, CNTs can also provide great support for nucleation and growth of ceramics such as HA and/or other materials [66].

The ability of CNTs to be functionalized has raised this material as a possible alternative to improve properties of biocomposites [68], [69]. This process consists in the adsorption of functional groups or molecules onto CNTs sidewalls without further changing on the structure of the nanotubes. According to Lacerda and co-workers [70], pristine CNTs are insoluble in most solvents and become cytotoxic due to this insolubility. The chemical process of functionalization makes them soluble improving their biocompatibility properties. In addition, bioactive agents can be conjugated to CNTs throughout functionalization and therefore can act as carriers for drugs, antigens and gene delivery.

The development of new and efficient drug delivery systems is of fundamental importance to improve the pharmacological field. The ability to carry one or more therapeutic agents with recognition capacity, optical signals for imaging and/or specific targeting is of fundamental advantage, e.g., in the treatment of cancer and different types of infectious diseases. The possibility to functionalize CNTs with bioactive agents and drugs has hence raised CNTs as a promising and efficient tool for transporting and translocating therapeutic molecules [71], [72].

Bianco and co-workers [71] studied and confirmed the solubility of functionalized CNTs in physiological media. Besides, they discovered that properly functionalized CNTs seem to have a high propensity to cross cell membranes. In addition, the chemistry of CNTs offers the possibility to introduce more than one function on the same tube wall, so that targeting molecules, contrast agents, drugs, or reporter molecules can be used at the same time. Although not yet commercialized, CNTs-based drug delivery systems are patented at the United States [73], [74].

Another reason for the use of functionalized instead of pristine CNTs is scaffolds. The presence of functional groups, e.g., may attract calcium cations to nucleate crystals of biocompatible materials, like HA [66].

Another recently important application of CNTs is in the manufacture of biosensors. A biosensor is a device that incorporates a biological sensing element, such as an enzyme, antibody, antigen, etc. which is associated with a physiochemical transducer. When an analyte is presented to the transducer, a chemical reaction takes place and provides an electrical signal that is proportional to the concentration of the analyte.

Up to date biosensors can be acquired in the marketplace by several companies, such as Biosensors International [75], BioDot Inc [76], Ercon Inc [77], etc. Their biosensors are produced with carbon, silver, platinum, etc. materials. Recently, the usage of nanomaterials, nonetheless, has developed a great interest on such field of applications; in particular one can cite CNTs as an alternative due to their chemical, electrical properties and mechanical strength.

The electrical property of CNTs prompts the electron transfer improving the sensitivity of the sensor. The ability to be functionalized also plays an important role in biosensors given that a wide range of drugs and molecules can be adsorbed onto their walls. Thus, CNTs-based biosensors are being studied for monitoring total cholesterol [78] and glucose [79], [80]. Most of researchers suggest that the glucose biosensor is the most prevalent biosensor in the marketplace today.

CNTs-based biosensors are not yet commercialized, although these materials are in research stage, one can find patents where CNTs are used in glucose sensors [81], [82].

6.3.2 Others

Another application of composites with great biomedical potential is the use of HA in combination with some polymers and other compounds as composites. Although these composites are not yet commercialized, the possibility of biomaterials with enhanced mechanical and biological proprieties is of great interest to scientists and biomaterials companies.

Many researchers have developed different types of nanocomposites involving HA. Among various combinations, some nanocomposites of great relevance are listed below:

- HA / chitosan [83];
- HA / polyanhydride [84];
- HA / poly(L-lactide) [85];
- HA / gelatin [86];
- HA / TiO_2 [87];
- HA / carbon nanotube [88].

6.4 Polymers

Some polymers are also studied and applied in the field of biomedicine, among them one can mention their usage as nanocarriers in the current antiretroviral drugs (ARVs), as e.g., to effectively reduce the Human Immunodeficiency Virus (HIV) viral load in the brain [89]. Polymeric nanoparticles such as Poly (butyl cyanoacryalate) (PBCA), methylmethacrylate-sulfopropylmethacrylate (MMASPM), poly (D, L-lactide-co-glycolide) (PLGA) and polylactide (PLA) increase the local drug concentration gradients, facilitating the drug transport into the brain.

6.5 Toxicity

Despite the properties that make engineered nanomaterials unique and beneficial, the idea that they may be harmful to humans has given rise to doubts about their risk-benefit [90]. Following this idea, it is increasingly recognized that the development of a nanomaterial to be used in future bio-applications should be accompanied by parallel studies to evaluate and understand their potential risk to the human health [91]. The European Agency for Safety and Health at Work has published a series of expert forecasts providing an overview of the potential emerging risks (physical, biological, psychosocial and chemical) of nanomaterials.

As previously mentioned, nanoparticles used in bio-applications can be similar in size to proteins, and are considerably smaller than many cells in the body. Thus, the human contact with nanoparticles can take place by several ways. Nanomaterials can be inhaled with air, swallowed, and may possibly enter the body via the skin. Due to the different ways of contact with nanoparticles, an exhaustive knowledge of the possible entry routes into the human body is necessary.

Once in contact with human body, nanoparticles can affect the lungs, circulation, genotoxic, and possible carcinogenic effects as well as some brief remarks on the brain can also occur. It is also possible that in some cases, nanoparticles can disrupt the biological processes, causing impairments, e.g., to structural or metabolic processes [90]. The communication between neighboring cells can also be influenced, because the nanoparticles can intercept or interrupt the exchange between enzymes. Another important risk of nanoparticles is the release of this material during the products' life cycle, which can affect the health of the consumer.

The report "Workplace exposure to nanoparticles", commissioned by The European Agency for Safety and Health at Work, concludes that there are many information deficits requiring research activities [92]. Moreover, at present there is limited data on the genotoxic potential of nanomaterials and their possible long-term impact on human health [93].

One of the major concerns of the use of CNTs-based biomaterials is the lack of knowledge on their effect to human health. The majority of papers published have suggested that pristine CNTs are cytotoxic. Toxicological studies of CNTs have proved, however, that this issue can be bypassed by the functionalization of nanotubes, as mentioned above. Nevertheless, some researchers claim that insufficient research was carried out so as to exclude the cytotoxicity of funcionalized CNTs [94]. Moreover, the chemical process of functionalization is not a solution when CNTs are inhaled. Up to date, several researcher are studying and exploring the toxic potential of CNTs to the lungs, and they claim that further evaluation of risk/benefit ratio is required [95], [96].

Furthermore, research results need to be further transcribed into standards and guidelines in different areas, such as exposure measurement or toxicological testing. At the moment, there are no promulgated government occupational exposure standards for nanomaterials. To cover the lack of standards, several government organizations have been working to fund and assemble toxicology information on these materials. The scientific community, government, companies, etc have discussed nanotoxicology issues as the need for detailed materials characterization; need for realistic exposure scenarios as well the need for methods to track nanomaterials in biological experiments.

References

[1] Wilson J, Hench LL (1993) An Introduction to Bioceramics. World Scientific Publishing Co., Singapura

[2] LeGeros RZ, LeGeros JP (1990) Dense Hydroxyapatite. In: Yamamuro T, Hench LL, Wilson J (eds) Handbook of Bioactive Ceramics, vol. II. CRC Press, Boca Raton

[3] Aoki H (1991) Science and Medical Application of Hydroxyapatite. JAAS, Tokyo.

[4] Natarajan UV, Rajeswari S (2008) Influence of calcium precursors on the morphology and crystallinity of sol–gel-derived hydroxyapatite nanoparticles. J Cryst Growth 310:4601–4611
[5] Simona V, Lazar D, Turcu RVF, Mocuta H, Magyari K, Prinz M, Neumann M, Simona S (2009) Atomic environment in sol–gel derived nanocrystalline hydroxyapatite. Mater Sci Eng B 165:247–251
[6] Ferraz MP, Monteiro FJ, Manuel CM (2004) Hydroxyapatite nanoparticles: A review of preparation methodologies. J Appl Biomater & Biomech 2:74-80
[7] Wang M (2004) Calcium Phosphate Ceramics. In: Shi D (ed) Biomaterials and Tissue Engineering. Springer, Berlin
[8] LeGeros RZ, LeGeros JP (1993) Dense Hydroxyapatite. In: Hench, LL, Wilson J (eds) An Introduction to Bioceramics. World Scientific Publishing Co., Singapura
[9] Zhang Z, Yang Y, Ong JL (2009) Nanohydroxyapatite for Biomedical Applications. In: Reisner DE (ed) Bionanotechnology – Global Prospects. CRC Press, Boca Raton
[10] Gross KA, Berndt CC (1997) Thermal Processing of Hydroxyapatite for Coating Production. J Biomed Mater Res 39:580–587
[11] Mobasherpour I., Heshajin MS, Kazemzadeha A, Zakeri M (2007) Synthesis of nanocrystalline hydroxyapatite by using precipitation method. J Alloy Compd 430:330–333
[12] Raynaud S, Champion E, Bernache-Assollant D, Thomas P (2002) Calcium phosphate apatites with variable Ca/P atomic ratio I. Synthesis, characterisation and thermal stability of powders. Biomater 23:1065–1072
[13] Liu H, Yazici H, Ergun C, Webster TJ, Bermek H (2008) An in vitro evaluation of the Ca/P ratio for the cytocompatibility of nano-to-micron particulate calcium phosphates for bone regeneration. Acta Biomater 4:1472–1479
[14] Hsieh M-F, Perng L-H, Chin T-S, Perng H-G (2001) Phase purity of sol–gel-derived hydroxyapatite ceramic. Biomater, 22:2601 – 2607
[15] Eshtiagh-Hosseini H, Housaindokht M R, Chahkandi M (2007) Effects of parameters of sol–gel process on the phase evolution of sol–gel-derived hydroxyapatite. Mater Chem Phys 106:310–316
[16] Natarajan UV, Rajeswari S (2008) Influence of calcium precursors on the morphology and crystallinity of sol–gel-derived hydroxyapatite nanoparticles. J Cryst Growth 310:4601–4611
[17] Liu D-M, Yang Q, Troczynski T, Tseng WJ (2002) Structural evolution of sol–gel-derived hydroxyapatite. Biomater 23:1679–1687
[18] Liu D-M, Troczynski T, Tseng WJ (2001) Water-based sol–gel synthesis of hydroxyapatite: process development. Biomater 22:1721 – 1730
[19] Liu D-M, Troczynski T, Tseng WJ (2002) Aging effect on the phase evolution of water-based sol–gel hydroxyapatite. Biomater 23:1227–1236
[20] Qiu C, Xiao X, Liu R (2008) Biomimetic synthesis of spherical nano-hydroxyapatite in the presence of polyethylene glycol Ceram Int 34:1747–1751
[21] Bernard L, Freche M, Lacout JL, Biscans B (1999) Preparation of hydroxyapatite by neutralization at low temperature: influence of purity of the raw material. Powder Technol 103:19-25
[22] Hayek E, Newesely H (1963) Pentacalcium monohydroxyorthophosphate. Inorg Synth 7:63-65
[23] Pang YX, Bao X (2003) Influence of temperature, ripening time and calcination on the morphology and crystallinity of hydroxyapatite nanoparticles. J Eur Ceram Soc 23:1697–1704

[24] Liu, D-M, Yang Q.; Troczynski T (2002) Sol–gel hydroxyapatite coatings on stainless steel substrates. Biomater 23:691–698
[25] Han JY, Yu ZT, Zhou L (2008) Hydroxyapatite/titania composite bioactivity coating processed by sol–gel method. Appl Surf Sci 255:455–458
[26] Jillavenkatesa A, Condrate Sr RA (1998) Sol–gel processing of hydroxyapatite. J Mater Sci 33:4111-4119
[27] Kokubo T (1990) Surface chemistry of bioactive glass ceramics. J Non-Cryst Solids 120:138-51
[28] Zhu W, Zhang X, Wang D, Lu W, Ou Y, Han Y, Zhou K, Liu H, Fen W, Peng L, He C, Zeng Y (2010) Experimental study on the conduction function of nano-hydroxyapatite artificial bone. Micro Nano Lett 5:19-27
[29] http://www.hydroxyapatite.com/ Accessed: 20th june, 2010
[30] http://www.fluidinova.com/ Accessed: 20th june, 2010
[31] http://www.sigmaaldrich.com/ Accessed: 20th june, 2010
[32] Karlsson C, Lundberg PJ, Rosinski A, Söderbom M (2003) Porous hydroxyapatite particles as carriers for drug substances. United States Patent 6558703
[33] Génin FY, Luo, P, Dash AK (2004) Hydroxyapatite based drug delivery implant for cancer treatment. United States Patent 6767550
[34] Flor J, Davolos, MR, Correa, MA (2007) Protetores solares. Quimica Nova 30: 153-158
[35] Serpone N, Dondi D, Albini A (2007) Inorganic and organic UV filters: Their role and efficacy in sunscreens and suncare products. Inorg Chim Acta 360:794–802.
[36] Araujo TS, Souza SO (2008) Protetores solares e os efeitos da radiação ultravioleta. Scientia Plena 4:1-7
[37] http://www.nanoamor.com/home Accessed: 13th june, 2010
[38] http://www.sbczn.co.kr/ Accessed: 29th june, 2010
[39] http://www.nanophase.com/ Accessed: 29th june, 2010
[40] http://www.micronisers.com/ Accessed: 5th june, 2010
[41] http://corporate.evonik.com/en/Pages/default.aspx Accessed: 11th june, 2010
[42] Trommer RM, Alves AK, Bergmann CP (2010) Synthesis, characterization and photocatalytic property of flame sprayed zinc oxide nanoparticles. J Alloy Compd 491:296–300
[43] Yamamoto O (2001) Influence of particle size on the antibacterial activity of zinc oxide. Int J Inorg Mater 3:643–646
[44] Rajendran R, et al (2010) Use of zinc oxide nano particles for production of antimicrobial textiles. Int J Eng Sci 2:202-208
[45] Gao J, Xu B (2009) Applications of nanomaterials inside cells. Nano Today 4: 37-51
[46] Shubayev VI, Pisanic II TR, Jin S (2009) Magnetic nanoparticles for theragnostics. Adv Drug Delivery Rev 61:467–477
[47] Sun C, Lee JSH, Zhang M (2008) Magnetic nanoparticles in MR imaging and drug delivery. Adv Drug Delivery 60:1252–1265
[48] http://berlex.bayerhealthcare.com/html/products/pi/Feridex_PI.pdf?WT.mc_id=www.berlex.com Accessed: 5th june, 2010
[49] Jain TK, Richey J, Strand M, Leslie-Pelecky DL, Flask CA, Labhasetwar V (2008) Magnetic nanoparticles with dual functional properties: Drug delivery and magnetic resonance imaging. Biomaterials 29:4012–4021
[50] Gupta AK, Gupta M (2005) Synthesis and surface engineering of iron oxide nanoparticles for biomedical applications. Biomaterials 26:3995–4021
[51] http://www.biophan.com Accessed: 4th july, 2010

[52] Roy I, Ohulchanskyy TY, Pudavar HE, Bergey EJ, Oseroff AR, Morgan J, Dougherty TJ, Prasad PN (2003) Ceramic-based nanoparticles entrapping water-insoluble photosensitizing anticancer drugs: a novel drug-carrier system for photodynamic therapy. J Am Chem Soc 125:7860-7865
[53] Sharma RK, Das S, Maitra A (2004) Surface modified ormosil nanoparticles. J Colloid Interf Sci 277:342–346
[54] Tripathi VS, Kandimalla VB, Ju H (2006) Preparation of ormosil and its applications in the immobilizing biomolecules. Sensor Actuat B chem 114:1071–1082
[55] Salata OV (2004) Applications of nanoparticles in biology and medicine. J Nanobiotechnology doi:10.1186/1477-3155-2-3
[56] Rai M, Yadav A, Gade A (2009) Silver nanoparticles as a new generation of antimicrobials. Biotechnol Adv 27:76–83
[57] Jain PK, El-Sayed IH, El-Sayed MA (2007) Au Nanoparticles target cancer. Nanotoday 2:18-29
[58] Partha G, Gang H, Mrinmoy D, Chae KK, Vincent M, Rotello (2008) Gold nanoparticles in delivery applications. Adv Drug Delivery 60:1307–1315
[59] Chen Y, Zhang YQ, Zhang TH, Gan CH, Zheng CY, Yu G (2006) Carbon nanotube reinforced hydroxyapatite composite coatings produced through laser surface alloying. Carbon 44:37-45
[60] Kaya C (2008) Electrophoretic deposition of carbon nanotube-reinforced hydroxyapatite bioactive layers on Ti–6Al–4V alloys for biomedical applications. Ceram Intern 34:1843-1847
[61] Balani K, Chen Y, Harimkar SP, Dahotre NB, Agarwal A (2007) Tribological behavior of plasma-sprayed carbon nanotubes-reinforced hydroxyapatite coating in physiological solution. Acta Biomater 3:944-951
[62] Hahn BD, Lee JM, Park DS, et al. (2009) Mechanical and in vitro biological performances of hydroxyapatite–carbon nanotube composite coatings deposited on Ti by aerosol deposition. Acta Biomater 5:3205-3214
[63] Li A, Sun K, Dong W, Zhao D (2007). Mechanical properties, microstructure and histocompatibility of MWCNTs/HA biocomposites. Mat Lett 61:1839-1844
[64] Li H, Wang L, Liang C, Wang Z, Zhao W (2010). Dispersion of carbon nanotubes in hydroxyapaptite powder by in situ chemical vapor deposition. Mat Scie & Eng B 166:19-23
[65] Xu JL, Khor KA, Sui JJ, Chen WN (2009) Preparation and characterization of a novel hydroxyapatite/carbon nanotubes composite and its interaction with osteoblast-like cells. Mat Scie and Eng C 29:44-49
[66] Harrison BS, Atala A (2007) Carbon nanotube applications for tissue engineering. Biomater 28:344-353
[67] Tran PA, Zhang L, Webster TJ (2009) Carbon nanofibers and carbon nanotubes in regenerative medicine. Adv Drug Deliv Rev 61:1097-1114
[68] Aryal S, Bhattarai SR, Bahadur R, et al. (2006) Carbon nanotubes assisted biomimetic synthesis of hydroxyapatite from simulated body fluid. Mat Scie & Eng A 426:202-207
[69] Liao S, Xu G, Wang W, et al. (2007) Self-assembly of nano-hydroxyapatite on multiwalled carbon nanotubes. Acta Biomater 3:669-675
[70] Lacerda L, Bianco A, Prato M, Kostarelos K (2006) Carbon nanotubes as nanomedicines: from toxicology to pharmacology. Adv Drug Deliv Rev 58:1460-1470
[71] Bianco A, Kostarelos k, Prato M (2005) Applications of carbon nanotubes in drug delivery. Curr Opin Chem Biol 9:674-679

[72] Simon F, Peterlik H, Pfeiffer R, Bernardi J, Kusmany H (2007) Fullerene release from the inside of carbon nanotubes: A possible route toward drug delivery. Chem Phys Lett 445:288-292

[73] Hirsch A, Sagman U, Rosenblum MG, Wilson LJ. Use of carbon nanotubes for drug delivery. United States patent US20080193490A1

[74] Tour JM, Moore VC, Casscella SW, et al. Water-soluble carbon nanotubes compositions for drug delivery and medicinal applications. United States patent US20090170768A1

[75] http://www.biosensors.com/intl/ Accessed: 25th may, 2010

[76] http://www.biodot.com/index.htm Accessed: 6th june, 2010

[77] http://www.erconinc.com/home.htm Accessed: 6th june, 2010

[78] Li G, Liao JM, Hu GQ, Ma NZ, Wu PJ (2005) Study of carbon nanotube modified biosensor for monitoring total cholesterol in blood. Biosensors & Bioelect 20: 2140-2144

[79] Sun TP (2010) Carbon nanotube composites for glucose biosensor incorporated with reverse iontophoresis function for noninvasive glucose monitoring. Intern J of Nanomedic 5:343-349

[80] Tsai MC, Tsai YC (2009) Adsorption of glucose oxidase at platinum-multiwalled carbon nanotube-alumina-coated silica nanocomposite for amperometric glucose biosensor. Sensors & Actuators B 141:592-598

[81] Lee J, Chung J, Lee K-H. Micro/Nano-fabricated glucose sensors using single-walled carbon nanotubes. United States patent US007118881B2

[82] Gu Y, Barrett EC. Carbon nanotubes-based glucose sensor. United States patent US 20050265914A1

[83] Chen F, Wang Z-C, Lin C-J (2002) Preparation and characterization of nano-sized hydroxyapatite particles and hydroxyapatite/chitosan nano-composite for use in biomedical materials. Mater Lett 57:858–861

[84] Haoying L, Chen Y, Xie Y (2003) Photo-crosslinking polymerization to prepare polyanhydride/needle-like hydroxyapatite biodegradable nanocomposite for orthopedic application. Mater Lett 57:2848–2854

[85] Hong Z, Zhang P, He C, Qiu X, Liu A, Chen L, Chen X, Jing X (2005) Nanocomposite of poly(L-lactide) and surface grafted hydroxyapatite: Mechanical properties and biocompatibility. Biomater 26:6296–6304

[86] Chang MC, Ko C-C, Douglas WH (2003) Preparation of hydroxyapatite-gelatin nanocomposite. Biomater 24:2853–2862

[87] Pushpakanth S, Srinivasan B, Sreedhar B, Sastry TP (2008) An in situ approach to prepare nanorods of titania-hydroxyapatite (TiO2–HAp) nanocomposite by microwave hydrothermal technique. Mater Chem Phys 107: 492–498

[88] Bhattarai SR, Aryal S, Bahadur R, Bhattarai N, Hwang PH, Yi HK, Kim HY (2008) Carbon nanotube-hydroxyapatite nanocomposite for DNA complexation. Mater Sci Eng C 28:64–69

[89] Wong HL, Chattopadhyay N, Wu XY, Bendayan R (2010) Nanotechnology applications for improved delivery of antiretroviral drugs to the brain. Adv Drug Deliv Rev 62:503–517

[90] Savolainen K, Pylkkänen L, Norppa H, Falck G, Lindberg H, Tuomi T, Vippola M, Alenius H, Hämeri K, Koivisto J, Brouwer D, Mark D, Bard D, Berges M, Jankowska E, Posniak M, Farmer P, Singh R, Krombach F, Bihari P, Kasper G, Seipenbusch M (2010) Nanotechnologies, engineered nanomaterials and occupational health and safety – A review. Saf Sci 48: 957-963

[91] Hurt RH, Monthioux M, Kane A (2006) Toxicology of carbon nanomaterials: Status, trends, and perspectives on the special issue. Carbon 44:1028–1033
[92] http://osha.europa.eu/en/publications/literature_reviews/workplace_exposure_to_nanoparticles/ Accessed: 12th july, 2010
[93] Singh N, Manshian B, Jenkins GJS, Griffiths SM, Williams PM, Maffeis TGG, Wright CJ, Doak SH (2009) NanoGenotoxicology: The DNA damaging potential of engineered nanomaterials. Biomaterials 30:3891–3914
[94] Firme CP, Bandaru PR (2010) Toxicity issue in the application of carbon nanotubes to biological systems. Nanomedic 6:245-256
[95] Shvedova AA, Kisin ER, Porter D et al. (2009) Mechanisms of pulmonary toxicity and medical applications of carbon nanotubes: Two faces of Janus? Pharmac & Therapeut 121:192-204
[96] Muller J, Huaux F, Lison D (2006) Respiratory toxicity of carbon nanotubes: How worried should we be? Carbon 44:1048-1056

Abbreviations

HA – hydroxyapatite
SEM - Scanning Electron Microscopy
SBF – synthetic body fluids
MNPs – magnetic nanoparticles
CNTs – carbon nanotubes
SPR - surface Plasmon resonance
HIV - Human Immunodeficiency Virus
ARVs - antiretroviral drugs
MRI - magnetic resonance imaging
UVA - ultraviolet A
PBCA - Poly (butyl cyanoacryalate)
MMASPM - methylmethacrylate-sulfopropylmethacrylate
PLGA - poly (D, L-lactide-co-glycolide)
PLA – polylactide

7 Nanomaterials and Catalysis

Annelise Kopp Alves[1], Felipe Amorim Berutti[2],
and Felipe Antonio Lucca Sánchez[1]

[1] Universidade Federal do Rio Grande do Sul, 90035190, Porto Alegre, Brazil
[2] Universidade Federal do Pampa, 96413170, Bagé, Brazil

Abstract. Catalysts are typical nanomaterials, perhaps the first nanomaterials in wide applications. Catalysis is a nanoscale phenomenon that has been the subject of research and development for many years, but only recently has it become a nanoscale science of materials and chemistry involving more investigations on the molecular level. Nanomaterial-based catalysts are usually heterogeneous catalysts. The extremely small size of the particles maximizes surface area exposed to the reactant, allowing more reactions to occur. However, thermal stability of these nanomaterials is limited by their critical sizes; the smaller the crystallite size, the lower the thermal stability. In this chapter the characterization of metal oxides such as CeO_2, TiO_2, and ZnO and some of their applications as catalysts for methane combustion and photocatalysis is described. The effects of mixed oxides, and mixed phases were investigated.

Keywords: catalytic combustion, photocatalysis, cerium oxide, titanium oxide, zinc oxide.

7.1 Introduction

Catalysts are typical nanomaterials, perhaps the first nanomaterials in wide applications. Catalysis is a nanoscale phenomenon that has been the subject of research and development for many years, but only recently has it become a nanoscale science of materials and chemistry involving more investigations on the molecular level.

Nanomaterial-based catalysts are usually heterogeneous catalysts. The extremely small size of the particles maximizes surface area exposed to the reactant, allowing more reactions to occur. However, thermal stability of these nanomaterials is limited by their critical sizes; the smaller the crystallite size, the lower the thermal stability.

Here, the characterization of metal oxides such as CeO_2, TiO_2, and ZnO and some of their applications as catalysts for methane combustion and photocatalysis is described. The effects of mixed oxides and mixed phases were investigated.

7.2 Catalytic Combustion

Ceria has been included in automotive emissions-control catalysts for many years and ceria-supported metals are finding new applications in hydrocarbon-reforming and in hydrocarbonoxidation catalysis [1].

In these applications, the ability of cerium to change its valence between Ce^{+3} and Ce^{+4} is decisive; however, the mechanisms by which ceria promotes reactions are not fully comprehended [2]. The prominent role of ceria has been recognized in three-way catalysis, catalytic wet oxidation, water-gas-shift reaction, oxidation/combustion catalysis and solid oxide fuel cells [3].

Ceria can be employed as an oxide carrier or a mixed oxide carrier with a transition metal oxide providing unique catalytic properties. Ceria is also believed to help in preserving the catalyst surface area, pore size distribution and catalytic activity. Several CeO_2-based systems such as CeO_2-ZrO_2, CeO_2-Al_2O_3, $CuO/CeO_2/Al_2O_3$, CeO_2-SiO_2, Pd/CeO_2 and Au/CeO_2 have been examined for their catalytic properties. Recently, efforts have also been made to synthesize nanostructured ceria particles with better physicochemical properties for diverse applications.

7.2.1 Structural Features of CeO_2

The structural properties of CeO_2 have been investigated by several authors providing valuable information on redox properties and oxygen mobility in the ceria lattice [4-6].

Ceria is a pale yellow color solid known to crystallize in fluorite structure (CaF_2) with a space group of Fm3m [7]. The unit cell of ceria is shown in Figure 7.1. In the face centered cubic (FCC) structure of ceria, Ce^{4+} ions form a cubic close packed arrangement and all the tetrahedral sites are occupied by the oxide ions whereas the octahedral sites remain vacant.

The fluorite structure of ceria is retained up to 900 K under reducing atmosphere. However, the lattice parameter is found to increase with reduction temperature, indicating an expansion in the FCC lattice [4]. The increase in the lattice parameter is attributed to the reduction of Ce^{4+} ions to Ce^{3+}. The radius of Ce^{3+} is larger than the radius of Ce^{4+} resulting in the lattice expansion. The ability of the cerium ion to switch between the Ce^{4+} and Ce^{3+} oxidation states depending on the ambient oxygen partial pressure is represented as:

$$CeO_2 \Leftrightarrow CeO_{2-x} + \frac{x}{2} O_2 \quad (7.1)$$

The amount of oxygen released in this reaction and the oxygen consumed in the reverse reaction is generally referred to as the oxygen storage capacity (OSC) of ceria material.

Under reducing atmosphere, ceria is known to form nonstoichiometric oxides of composition CeO_{2-x} where $0<x<0.5$. The oxidation of CeO_{2-x} occurs at room temperature, while the reduction of CeO_2 starts at 473K. Pure CeO_2 deactivates its OSC usually at 1123K due to its sintering [8].

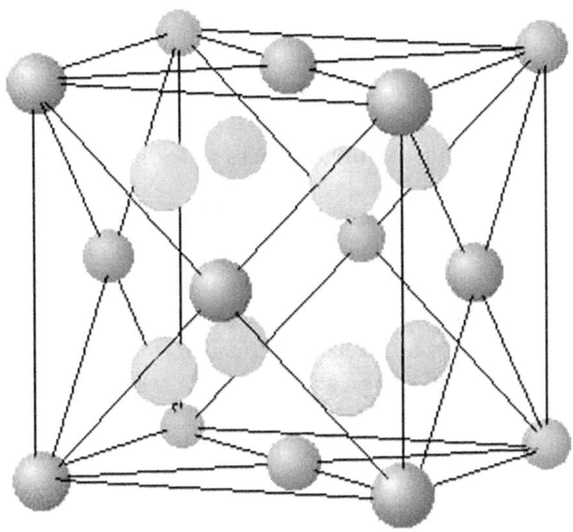

Fig. 7.1. Fluorite structure of CeO_2.

High OSC is believed to be dependent on a high surface area and the phase formed during synthesis. Ceria materials with high OSC are also obtained by cation doping, which produces oxygen vacancies [9].

Ceria in Three Way Catalyst (TWC)

The main application of the OSC of ceria-based materials is in exhaust catalysis, extensively reviewed in the literature [10-15]. The three-way catalysts are designed to convert automobile exhaust pollutants such as uncombusted hydrocarbons (HC), carbon monoxide (CO) and nitrogen oxides (NOx), to environmentally acceptable products such as carbon dioxide, water and nitrogen.

The TWC formulation usually contains metals (Pt, Rh, and Pd) dispersed on the surface of alumina. High surface area γ-alumina is used as support because of its higher thermal stability under hydrothermal conditions. Ceria and ceria-based mixed oxides are used as oxygen storage promoters in TWC. An optimum performance of the TWC is possible if the stoichiometric air–fuel ratio (A/F) is maintained at about 14.6 [16].

If the air–fuel ratio is close to the stoichiometric value of 14.6, the catalyst converts all the pollutants to CO_2, H_2O and N_2 gases with high efficiency. Under fuel lean periods the conversion of NO is affected whereas under fuel-rich conditions the oxidation of CO and hydrocarbons remains incomplete. The role of ceria and ceria-based materials in TWC is to widen the A/F window and help maintaining the conversion efficiency of the catalyst.

Application of CeO_2 Nanostructured Fibers as Catalyst

Due to more restrictive environmental regulations, nowadays the majority of automobiles with internal combustion engines have a system that controls the emission of exhaust gases: carbon monoxide (CO), hydrocarbon compounds (CxHy) and nitrogen oxides (NOx), called a catalytic converter.

The catalytic converter was invented by Eugene Houdry around 1950, when the results of early studies of smog in Los Angeles were published [17]. Houdry founded a company, Oxy-Catalyst, to develop catalytic converters for gasoline engines [18]. The catalytic converter was further developed by John J. Mooney and Carl D. Keith at the Engelhard Corporation, bought by BASF in 2006, creating the first three-way catalytic converter in 1979/80 [19].

A typical catalytic converter is based on CeO_2 doped with noble metals such as Pt and Pd. The efficiency of this converter depends on the activity of the catalyst. There are several industrial applications of nanostructured CeO_2 catalysts and almost every car manufacturer develops its own CeO_2 catalytic system [20-23].

In this chapter the development of catalysis tests, using nanostructured cerium oxide doped with copper fibers, is discussed. The fibers were obtained using the electrospinning process. To summarize, the electrospinning process uses a high power supply to generate an electrical field on which a polymer solution containing ions is accelerated and elongated, forming fibers of micro to nanoscale diameters. More details on the electrospinning process can be found in the literature [24-26].

The structure of the fibers produced using this technique can be seen in Figure 7.2a and b.

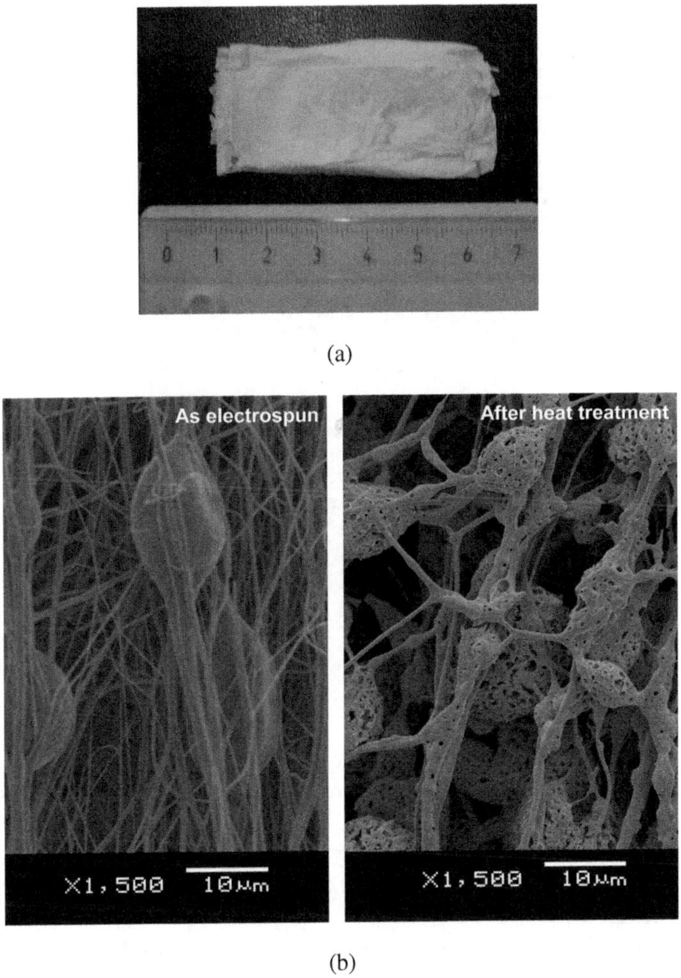

Fig. 7.2. (a) Photography of the CeO_2 fiber-mat obtained using the electrospinning process. (b) SEM image of the fiber mat showing the microstructure of the fibers.

The average size of the crystallites of the electrospun CeO_2 fibers is 25nm. Figure 7.3 presents a TEM showing a nanostructured fiber. One can notice that a single fiber contains several crystallites in nanoscale.

This fibrous material was used as a catalyst in the combustion of methane in air under different temperatures. The apparatus built to analyze the catalytic activity of the fibers consists of a vertical tubular furnace, a quartz tube reactor, a set of methane and air flow controllers, and a thermocouple for measuring the temperature of the reaction. The air and methane flow was kept constant at 0.9 L/min and 0.1 L/min, respectively. The amount of sample used was approximately 0.2 g.

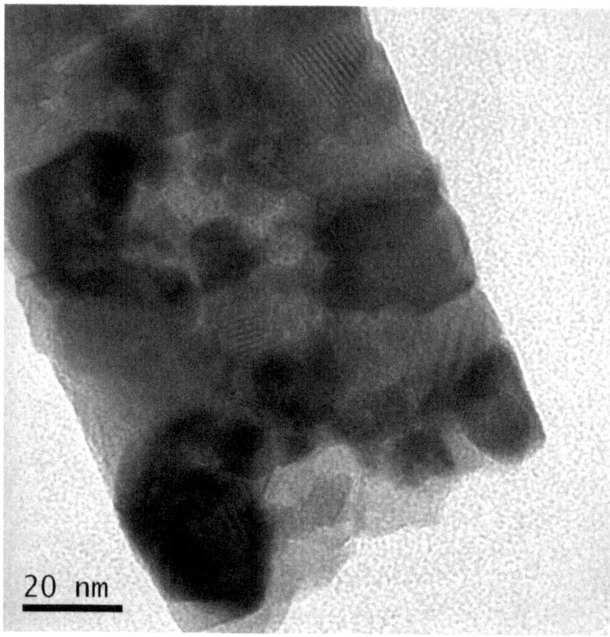

Fig. 7.3. TEM image of the nanostructured CeO_2 fibers obtained by electrospinning.

In a typical catalytic combustion test the sample is loaded into the quartz reactor, and set into the furnace. The gas mixture inlet is connected to the quartz reactor. The temperature of the furnace is set to reach 600°C in 60 minutes with a dwell time at 600°C of 5 minutes. A gas analyzer ECOLINE 4000 was used to analyze the quantities of O_2, CO, CO_2, NO, NOx and CxHy, consumed or generated during the reaction time.

The results of the catalysis tests using cerium oxide doped with 0.5, 1 and 2,5% are presented in Figure 7.4 in terms of the conversion of methane. The ignition temperature and the total methane conversion using the different amounts of copper are presented in Table 7.1.

Table 7.1. Composition of the catalyst ignition temperature and methane conversion.

Catalyst	**Ignition Temperature (°C)**	**Methane Conversion (%)**
CeO_2	594	85.4
CeO_2 + 0.5% Cu	559	100.0
CeO_2 + 1.0% Cu	576	91.0
CeO_2 + 2.5% Cu	593	71.2

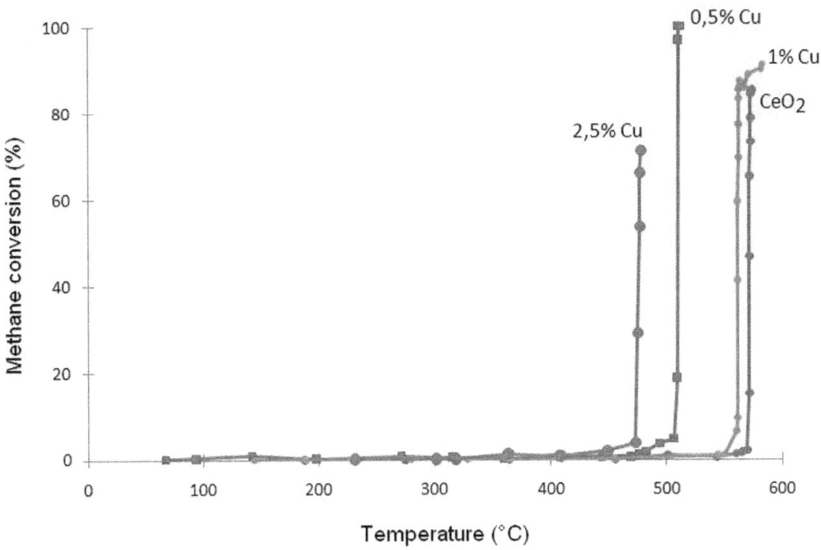

Fig. 7.4. Conversion of methane during the catalysis tests using cerium oxide doped with 0.5, 1 and 2,5%.

One can observe that the addition of more than 1% of copper has no beneficial effects over the reaction conversion. The presence of a metal such as copper over cerium oxide promotes the catalytic effect of the combustion reaction, but this effect is apparently limited to 1% of copper.

7.3 Photocatalysis

7.3.1 Introduction

Photocatalysis is a process where a chemical reaction is accelerated in the presence of a catalyst that is only active in the presence of light (ultraviolet or visible light). The oxidation of most hydrocarbons proceeds slowly in the absence of a catalytic active substance. A photocatalyst decreases the activation energy, making photoinduced processes occur.

A photocatalytic system usually consists of a semiconductor particle (photocatalyst) which is in close contact with a liquid or a gaseous reaction medium. When a semiconductor is exposed to UV light for example, redox reaction will occur at the surface of the catalyst generating hydroxyl radicals ($^{\bullet}OH$) and superoxide ions (O^{-2}). These oxidizing agents are able to decompose organic compounds into CO_2 and H_2O [27].

The most common photocatalysis applications are: conversion of water to hydrogen gas by photocatalytic water splitting; self-cleaning glasses, textiles and surfaces; disinfection of water; oxidation of organic contaminants using magnetic particles that are coated with photocatalyst nanoparticles; conversion of carbon dioxide into gaseous hydrocarbons; as an UV light absorber; sterilization of surgical instruments; among many other applications [28].

7.3.2 Titanium Oxide

Nanocrystalline photocatalytic materials are a potentially lucrative area of nanomaterials development [20]. Industrial utilization of the photocatalytic effect of nanomaterials is already seen in applications such as self-cleaning tiles, windows and textiles, anti-fogging car mirrors, and anti-microbial coatings [29-32]. For the photocatalysis market, nanostructured titanium dioxide (TiO_2) is the most suitable material for industrial use at present and promises innovative products in the future.

The commercial potential of nanoscale TiO_2 coatings is enormous, including applications in medicine, architecture (cultural heritage protective/recuperation purposes, facade paints, indoor, wall paper, tiles, etc.), the automotive and food industries (cleaner technologies, non-fogging glass and mirrors, product safety), the textile and glass industries, and in environmental protection (water and air purification and disinfection) [33-38].

Titanium dioxide nanoparticles (anatase and/or rutile phases) are able to generate excited ions by promoting electrons across theirs band gaps. The electrons and holes react with the surrounding water molecules to produce hydroxyl radicals and protons. These radicals mineralize organic compounds to CO_2 and H_2O. For pure titanium dioxide this effect only occurs under UV illuminating, while Visible light may produce this effect in doped titanium dioxide.

Application of TiO_2 Nanostructured Fibers as Photocatalyst

Commercial nanostructured TiO_2 photocatalysts are made by several different companies around the world. The most studied and published photocatalyst is made by Evonik Industries using the high-temperature flame hydrolysis; it is known as AEROXIDE® TiO_2 P25 [24]. P25 is commonly used as a standard photocatalyst and many researches are developing new materials to achieve its high photocatalysis efficiency.

In this chapter the photocatalysis tests were made with nanostructured titanium dioxide fibers and the results compared with TiO_2 P25. The fibers were obtained using the electrospinning process. To summarize, the electrospinning process uses a high power supply to generate an electrical field on which a polymer solution containing ions is accelerated and elongated, forming fibers of micro to nanoscale diameters. More details on the electrospinning process can be found in the literature [24-26].

The structure of the titanium dioxide fibers produced using this technique can be seen in Figure 7.5a, b and c.

(a)

(b) (c)

Fig. 7.5. (a) Photography of the TiO_2 fiber-mat obtained using the electrospinning process. (b) SEM image of the fiber mat showing the microstructure of the fibers. (c) High resolution FESEM image showing details of the microstructure of a single fiber.

The average size of the crystallites of the electrospun TiO_2 fibers is 20nm. Figure 7.6 presents a TEM showing a nanostructured fiber. One can notice that a single fiber contains several crystallites in nanoscale.

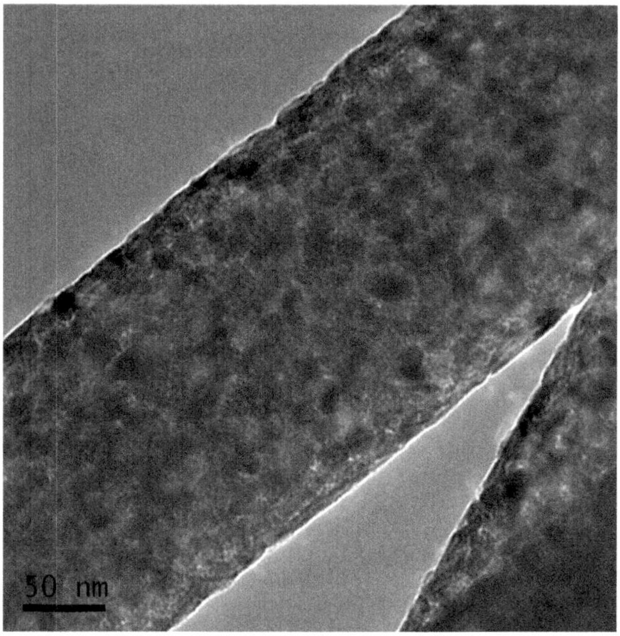

Fig. 7.6. TEM image of the nanostructured TiO_2 fibers obtained by electrospinning.

This fibrous material was used as a model photocatalyst in the decomposition of organic dyes, such as methylene blue (MB) and methyl orange (MO). The apparatus built to analyze the photocatalytic activity of the fibers consists of a Dreschel glass flask of 130mL equipped with a rubber septum to collect samples. This flask is located between two half cylinders, each with a set of 12 Ultra-Violet (black light) or Visible T8 type lamps. The system also contains a magnetic stirrer, an air bubbler and a thermostatic bath to keep the reaction temperature constant at 30°C. More construction details of this equipment can be found in the work of Alves, et al. [39].

A typical photocatalysis experiment consists of mixing 50mg of the catalyst with 125mL of a 20ppm dye solution. This mixture is homogenized with the help of an ultrasound finger for about 15 minutes. After this period a 4 mL sample of this solution is taken and poured into a PMMA spectrophotometer cuvette. The solution is transferred to the Dreschel flask and the air bubbler and the UV or visible lights are turned on. Samples are then taken every 5 minutes to follow the photodecomposition of the dye. After about 60 minutes of irradiation the samples are analyzed using a Uv–vis spectrophotometry.

The results of the photoactivity of the fiber in comparison with P25 are depicted in Figure 7.7 for the decomposition of methylene blue under UV illumination.

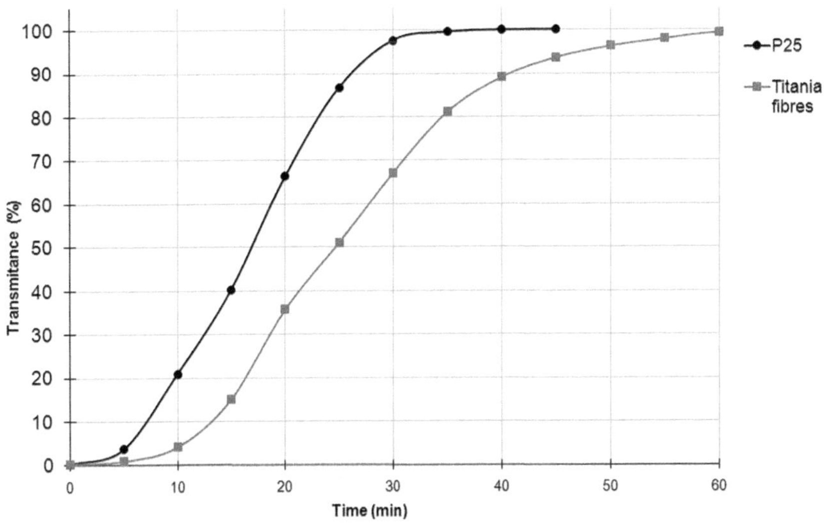

Fig. 7.7. Decomposition of methylene blue under UV light using TiO_2 nanostructured fibers and P25.

It is well known that the pH of the medium strongly influences the photocatalytic reactions. An analysis of the effect of the pH on the photoactivity of the nanostructured fibers was made using methyl orange as a dye model. Methyl orange was the selected model dye because of its high stability in the entire pH range. The results of these experiments are depicted in Figure 7.8.

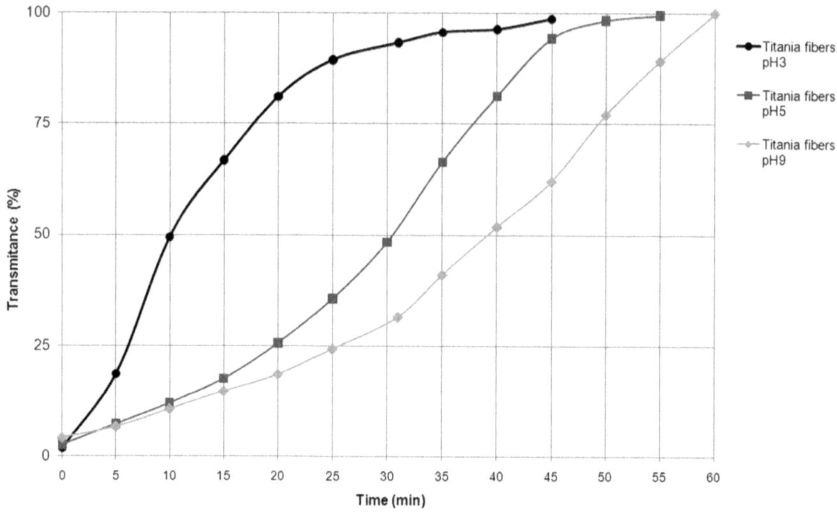

(a)

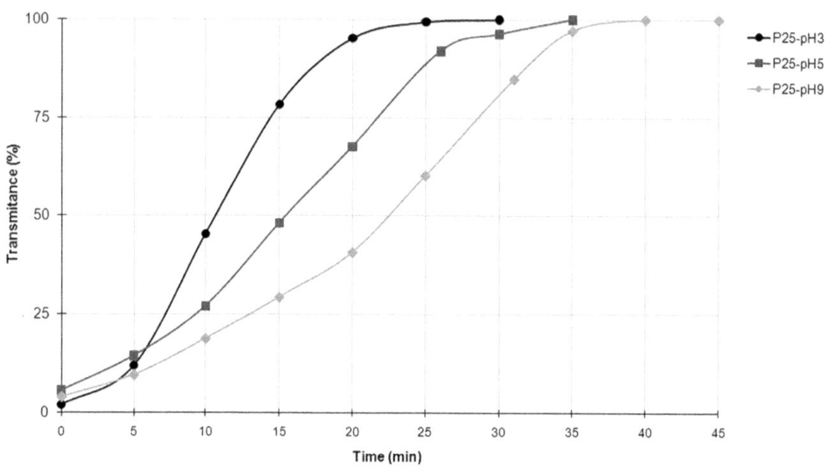

(b)

Fig. 7.8. Decomposition of methyl orange under UV light using (a) TiO_2 nanostructured fibers and (b) P25, under different pH condition.

Doping TiO_2 with other metals can promote the photocatalytic effect in visible light. During the preparation of the TiO_2 nanostructured fibers using electrospinning, different amounts of tin were added to study the contribution of this metal in the UV and Visible photoactivity of TiO_2. The results of these experiments under

UV light are depicted in Figure 7.9 and the results of the experiments under Visible light are depicted in Figure 7.10.

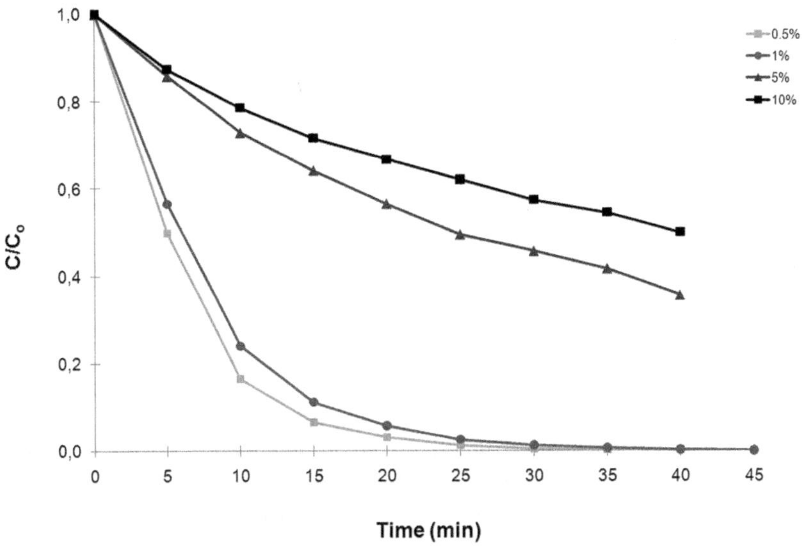

Fig. 7.9. Photodecomposition of methyl orange under UV light using TiO_2 nanostructured fibers doped with tin.

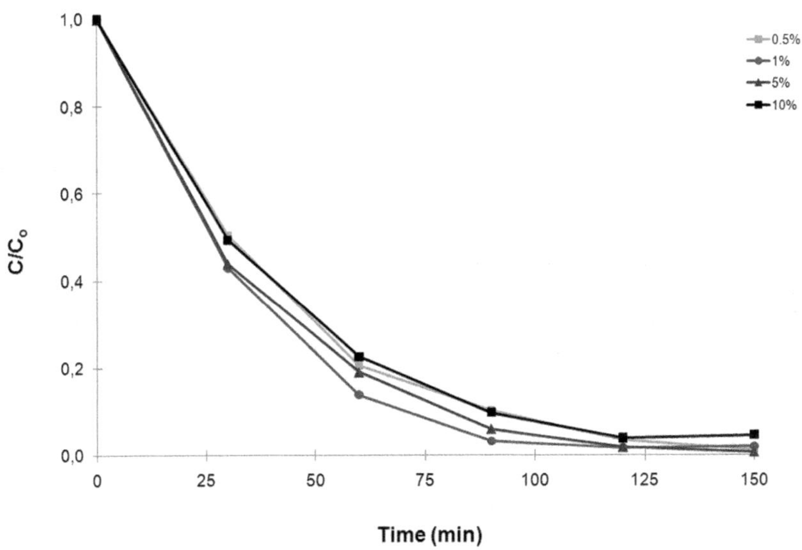

Fig. 7.10. Photodecomposition of methyl orange under Visible light using TiO_2 nanostructured fibers doped with tin.

Nanostructured fibers of TiO_2, with and without tin, obtained using the electrospinning technology was highly active in UV and Visible light.

7.3.3 Zinc Oxide

Zinc oxide is an inorganic compound with ZnO as chemical formulation. In high purity levels it is a white powdery substance, which is nearly insoluble in both water and alcohol, but soluble in most of the acids including hydrochloric acid. Due to its variety of interactive properties in distinct kinds of combined systems or compositions, zinc oxide has an extensive gamma of profitable uses in a wide range of industrial applications in different technological fields. Some well known and significant industrial applications of this versatile material are described below:

• Rubber Industry: The chemical reactivity of zinc oxide contributes to activate the organic accelerator, to speed up rubber cure rate. It is also useful to protect latex plantation because it reacts with the enzyme responsible for its decomposition. Due to its dielectric properties, zinc oxide improves the resistance to corona effects since its dielectric strength maintains a series of physical properties of rubber compounds at high operating temperatures and, similarly, because it retards devulcanization of many types of rubber compositions operating at elevated temperatures. Through its high brightness, refractive index, and optimum particle size, zinc oxide provides a high degree of whiteness and tinting strength for such rubber products as tire side walls, sheeting and surgical gloves.

• Ceramic Industry: Some of its newer applications are as electronic glass, low-melting glass for metal-to-glass seals, thermistors for use as lighting arresters and devitrified glasses of low thermal expansion. Zinc oxide plays an important role in semiconductor ceramic elements for operation at elevated temperatures or high voltages. Thermoelements such as varistors are composed of zinc oxide semiconductors modified or doped by other elements and can be produced to cover a broad range of thermal and electrical needs.

• Pharmaceutical and cosmetic Industries: Zinc oxide is mainly used in zinc soap, ointment, dental inlays, food powders, etc. Its optical and biochemical properties impart special features to a variety of cosmetic preparations for hair and skin care. In powders and creams it protects the skin by absorbing the ultraviolet sunlight rays and in burn ointments it aids healing.

• Plastics Industry: In plastics manufacturing, zinc oxide offers a large range of advantages such as mechanical and thermal resistance and it increases the lifetime of some polymer products by stabilizing them against aging and UV radiation as well.

• Paint Industry: Zinc oxide in organic coatings provides a broad spectrum of optical, chemical, biochemical and physical properties. Over the past century the paint industry, in its constant development of improved products, has utilized various aspects of those properties to a high degree.

• Adhesives Industry: Zinc oxide has long been a major constituent of surgical and industrial tapes based on natural or synthetic rubber as it is outstanding in retention of tack during shelf aging.

• Foods and Food-Packaging Industries: Zinc oxide and its derivatives contribute specially as fungistatic element and because of its chemical properties to the processing and packaging of various animal and vegetable products. It has long been incorporated into the varnish linings of the metal containers to prevent formation of black sulfides which discolor the food.

Other zinc oxide applications include its use in industrial lubrificants composition, in the photocopying process, in cigarette filters, for sulfur removal in fluids, in Portland cement composition and as a component of batteries, fuel cells and photocells.

This oxide occurs in the Earth's crust as a mineral, known as zincite and can be found in a few deposits all over the world. However, industrial production of zinc oxide is basically obtained by three general methods [40]:

• The French process: Also known as the indirect process, this technique starts by heating pure metallic zinc in an appropriate crucible until the boiling point (temperatures above 907 °C). The zinc vapor obtained naturally reacts with the oxygen in the air, resulting in a white, zinc oxide powder due to an oxidation reaction. The particles are transported into a cooling duct and collected in an adequate particle trap system. This indirect method was popularized by LeClaire (France) in 1844 and therefore is commonly known as the French process. Its product normally consists of agglomerated zinc oxide particles with an average size of 0.1 to a few micrometers. By weight, most of the world's zinc oxide is manufactured via the French process.

• The direct (American) process: In the direct process, the starting materials are various contaminated zinc composites, such as zinc ores or smelter by-products. It is reduced by heating with a carbon additive (e.g. anthracite) to produce zinc vapor. The generation of carbon monoxide reduces the oxidic zinc and the zinc is expelled as vapor. The zinc is reoxidized when it comes into contact with lower temperature air, forming zinc oxide particulate. The combination of the reduction reaction and the zinc vapor pressure combine to purify the zinc which is then oxidized as in the indirect process. Because of the lower purity of the source material, the final product is of lower quality in the direct process as compared to the indirect one.

• Wet chemical processes: These routes usually start with purified zinc solutions from which zinc carbonate or zinc hydroxide are precipitated. Before the final product is obtained, some stages are required, such as, filtration, washing, drying and calcination (mild temperatures ~ 800°C). For example, this zinc oxide is produced as a co-product with sodium hydrosulfite, a bleaching agent used primarily in the Paper Industry. Sulfur is present as zinc sulfide and other sulfur containing compounds, with total sulfur content reaching up to 2% and producing a lemon-yellow colored powder. Grades which are manufactured from precipitated zinc hydroxide or basic zinc carbonate intermediates are often termed as active zinc oxide. Mild heat treatment decomposes the materials into zinc oxide. This oxide is relatively reactive, due to its very high surface area which can be above 50 m^2/g. Such products are useful in catalysts and latex rubber applications.

Other controlled thermal treatments make possible the generation of suitable products for different markets. There is a very wide range of specifications for different end uses and geographical locations, how these are met depends on the practices of the individual zinc oxide manufacturer.

Zinc oxide production occupies an important place in the structure of the global zinc market. The annual global consumption of zinc is 10.5-11 million tons, and about 8% of it is used in the zinc oxide production. The largest consumption sector of this product is the Rubber Industry for the manufacture of tires and mechanical rubber goods [41].

Zinc oxide is a semiconductor with a large exciton binding energy (60 meV), direct band-gap (E_g = 3.37 eV) exhibiting near UV emission, transparent conductivity and piezoelectricity. Moreover, it is biosafe and biocompatible, and may be used for biomedical applications without coating.

Intensive research has been focused on the fabrication of zinc oxide nanostructures and on correlating their morphologies with their size-related optical and electrical properties [42-46]. Various kinds of zinc oxide nanostructures have been realized, such as nanobelts, nanodots, nanorods, nanowires, nanotubes, nanobridges, nanonails, nanowalls, nanohelixes and seamless nanorings among other varieties [47-49].

Strict parameter controls of conventional zinc oxide processes as well as some adaptations in the ordinary techniques are the key in the development of zinc oxide nanostructured powders, which will be referred to from this point as "nano-ZnO". Two different techniques are being developed by the authors and will be further detailed in this chapter.

The European Commission has listed nano-ZnO as one of the major nanomaterials to be intensely commercialized between 2006 and 2014. Nano-ZnO has attracted lots of basic research funding due to its highly versatile and promising applications in biotechnology (antibacterial, antifungal and antifouling), UV protection (sunscreens, paints, polymer nanocomposites and rubber), optoelectronics (LEDs, hydrogen fuel, toner, sensors), among other exposed applications [50].

The global production of zinc oxide equals over 1 million tons of zinc oxide per year [51], less than 1% of this, however, is nano-ZnO. Currently, the major production of nano-ZnO is mainly to supply the Sunscreen Industry where nano ZnO (together with TiO2) serves as the UV absorber. Around 30% of sunscreen lotion uses nano ZnO particles [52] as the photocatalysts that protects the human skin from dangerous UV rays of sunlight.

Due to the remarkable photochemical reactivity of nano-ZnO, it can match the reactivity of micrometric ZnO with far smaller amounts of nano-ZnO, and this approach, in turn, can greatly reduce the release of toxic zinc metal into the environment.

Most of the work on nano-ZnO is still at the basic research stage [50]. It is expected that by 2015, ZnO nanomaterials will be mass produced in a much larger tonnage for various applications in automotive catalysis, sunscreens, antibacterial products, rubber vulcanization, fungicides, pigments, nano-textiles, high-performance coatings, water treatment, dental fillings, hydrogen fuel systems, chemical sensors, pollutant filters, toner photoconductor, and optoelectronics [50, 53-54]. In any case, it is possible to find several industries spread around the world that are producing different kinds of nanostructurated zinc oxides on a large scale [55-61].

Photocatalytic Activity Applications of Nanostructured Zinc Oxide

Semiconductor materials such as zinc oxide show a vast number of interesting properties, which are maximized when these belong to the nanostructured materials as has already been discussed previously. However, one of the emerging and intensively explored properties of this nanostructured oxide is its photocatalytic activity mainly for the treatment of environmental pollution.

The Environmental Protection Agency (USA) has listed several volatile organic compounds as contaminants in ground water. The list includes several chlorinated aromatic and aliphatic compounds such as 4-chlorophenol, pentachlorophenol, chlorobenzene, dichlorobenzene, carbon tetrachloride, chloroform, dichloroethylene, trichloroethylene and dichloromethane, and other common organic solvents such as phenols, toluene, benzene and xylenes [62].

Established waste water treatments such as activated carbon adsorption, membrane filtration, ion exchange on synthetic adsorbent resins, chemical coagulation, etc., also generate wastes during the water treatment process, which requires additional steps and consequently more investment. In recent years, the heterogeneous photocatalytic oxidation (HPO) process employing ZnO and UV light has emerged as a promising new route for the degradation of persistent organic pollutants, and produces more biologically degradable and less toxic substances [63-64].

The photocatalytic phenomena of zinc oxide occur due to the presence of a large number of defects in the crystalline structure such as oxygen vacancies, interstitial zinc atom from the donor states, zinc vacancies and intestinal oxygen atoms from the acceptor states. The electronic condition created in this structural arrangement produces a band-gap, an intrinsic characteristic of semiconductor materials, by the existence of a forbidden gap between the valence and the conduction band. The photocatalytic activity of zinc oxide occurs only when photons with energies greater than the band-gap energy can result in the excitation of valence band electrons, which can then promote a reaction. The absorption of photons with lower energy than the band-gap energy or longer wavelengths, usually causes energy dissipation in the form of heat. The illumination of the photocatalytic surface with sufficient energy leads to the formation of a positive hole in the valence band and an electron in the conduction band [65]. After the light excitation the pair electron-hole created could have an electronic recombination releasing heat or could be involved in electronic transfer with other species and interact directly or through some photosensitizers with organic substrates. The positive hole oxidizes either pollutant directly or water to produce hydroxyl radicals (•OH), whereas the electron in the conduction band reduces the oxygen adsorbed on the catalyst [66]. Based on this principle zinc oxide and more recently nano-ZnO have received attention in research on the degradation of different organic specimens that are environmental pollutant.

An important feature in the environmental photocatalysis is the selection of semiconductor materials such as ZnO and TiO_2, which are close to being two of the ideal photocatalysts in several research proposals. These oxides are relatively inexpensive, and they provide photogenerated holes with high oxidizing power due to their wide band gap energy [67]. Since ZnO has almost the same band-gap energy as TiO_2, its photocatalytic capacity is anticipated to be similar to that of TiO_2. However, in the case of ZnO, photocorrosion frequently occurs with the illumination of UV light, and this phenomenon is considered one of the main reasons for the decrease of ZnO photocatalytic activity in aqueous solutions [68,69]. However, some studies have confirmed that ZnO exhibits a better efficiency than TiO_2 in photocatalytic degradation of some dyes, even in aqueous solution [70,71]. ZnO has sometimes been reported to be more efficient than TiO_2. Nevertheless, the biggest advantage of ZnO is that it absorbs over a larger fraction of the solar spectrum than TiO_2 [72]. For this reason, ZnO is quite suitable photocatalyst for photocatalytic removal in the presence of sunlight [73].

As was previously mentioned in this chapter, two researches that are being developed by the authors focused on the production of nano-ZnO powders will be described. This nanostructured material is being produced, on the one hand, by a suitable adaptation of the extensively used industrial technique (French process) and, on the other hand, based on a flame synthesis process.

The first nano-ZnO process mentioned is based on the thermal evaporation of pure metallic zinc in an oxidative atmosphere and is in development as an industrial prototype. It consists basically of a horizontal tube furnace with strict control of parameters such as temperature of the reactional environment, gas flow and the position of the injection points of the gases. The apparatus that was newly developed is passing through a patent request process. The methodology employed appears to be an interesting alternative for producing nano-ZnO on a large scale without any residual pollutants. Depending on the procedural conditions, the nano-ZnO produced can be either shaped as nanowires or nanotetrapods, with a white, agglomerated powder aspect.

The second technique developed for obtaining nano-ZnO consists of a flame synthesis. This flame-based equipment was also completely assembled in the laboratory. This method, basically, consists of forming a spray from a liquid combustible solution as the precursor, which is then burnt at the high temperatures of the flame, producing the nano-ZnO product, which is later collected in a powder collector system [74]. The nano-ZnO produced is composed of espherical primary nanoparticles that remain in an aggregated state. Flame-based techniques are a continuous and well-established method and are considered cost-effective processes that have the potential for large scale nanoparticles production [75,76].

The wide range of morphologies of nano-ZnO are well illustrated in the literature as well as the considerable amount of processes and applications that are being developed. It is also clear that each nanostructure created proves to have great effects on their widely varying properties and corresponding potential applications [77-80].

An interesting approach at this point is to compare the photocatalytic activity of both nano-ZnO materials synthesized by the authors through the mentionated procedural techniques and have an evident distinct morphology. Figure 7.11 shows micrographies produced by scanning electron and transmission electron microscopic techniques of both nano-ZnO powders. The thermal evaporation-based nano-ZnO (ZnO TEb) reveals a structural shape of tetrapods with an acicular edge diameter of 30–50 nm and a core diameter of a few micrometes. The flame synthesis-based nano-ZnO (ZnO-FSb) shows the primary particles with a spherical shape with 20–40 nm of diameter. The specific surface area of the nano-ZnO powders, measured by Brunauer–Emmett–Telle (BET) N_2 sorption technique, results in significantly different values corresponding to 8.35m²/g for ZnO-TEb and 17.34 m²/g for the ZnO-FSb. The discrepancy of the measurements can be attributed to the powder agglomeration that the nanotetrapod shaped ZnO acquired after being processed as well as to some metallic zinc that could remain after processing, which is being investigated.

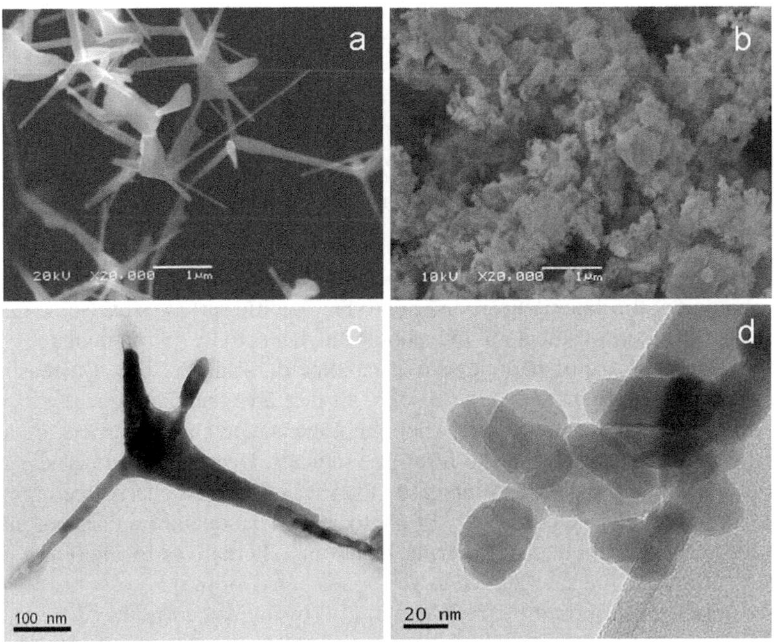

Fig. 7.11. Scanning electron migrographies of nano-ZnO obtained via the French process (a) and the Flame process (b). Transmission electron micrographies of nano ZnO obtained via the French process (c) and Flame process (d).

The photocatalyitic activity of the nano-ZnO was measured by the evaluation of a standard solution of methyl orange (MO) in a photochemical reactor using the same methodology as described in the work of Trommer R. M. et al [74]. Figure 7.12 shows the photocatalytic degradation of MO by the nano-ZnO obtained. One can observe that in a period of 20 min the ZnO-TEb reaches almost 80% of MO degradation while the ZnO-FSb achieves 10% of degradation.

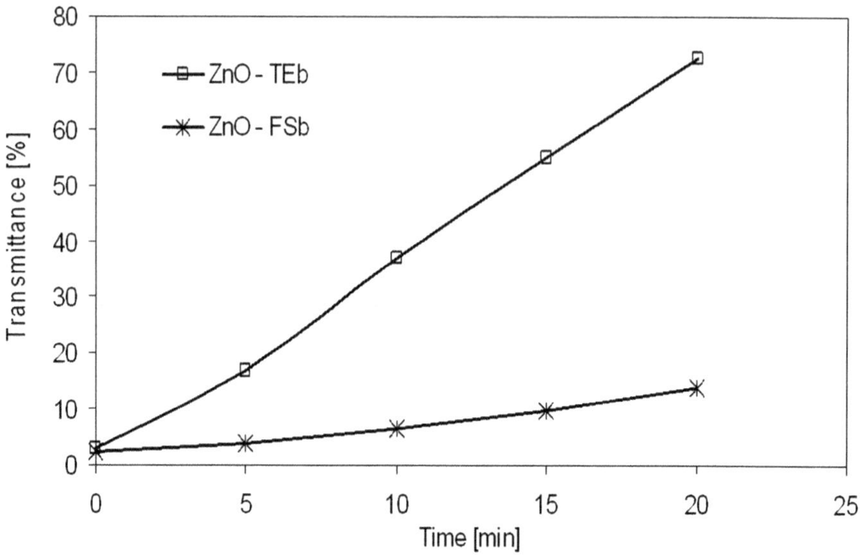

Fig. 7.12. Photocatalytic activity of nano-ZnO with different morphology described by a Transmittance versus time in the degradation of methyl orange dye.

The different behavior of the nano ZnO materials can be attributed to the morphology of the nanoparticles and not to the specific surface area. Some research discusses the improvement of the photocatalytic activity of particulate systems through the decrease of particle size at nanometric scale [81-83]. Decreasing the average particle size increases the specific surface area and consequently increases the number of active surface sites where photogenerated charge carriers are able to react with adsorbed molecules to form free radicals. Decreasing the particle size of a photocatalysts, though, also increases the rate of surface charge recombination [84]. As a consequence the activity of photocatalyst nanoparticles does not increase with decreasing particle size inferring that some alternatives to inhibiting charge carrier recombination are required to verify this behavior [85].

An argument proposed by Y. Wang et al [86] indicated that the differences in the photocatalytic activity of distinct nanostructures of zinc oxide could be associated with the number of oxygen vacancy sites present. These vacancies act as active center and are the main reason for causing the difference in photocatalytic activities for varied morphologic nano-ZnO.

7.4 Concluding Remarks

Summarizing: it is strictly necessary to understand the physiochemical principles involved and correlated with the photocatalytic behavior of species such as TiO_2 and ZnO. Comprehending the behavior of these materials in different environments such as water or air, as well as the radiation used to stimulate the photocatalytic activity and correlating these systems with the morphology and particle size of the material to be used, are fundamental to develop a guide for the treatment of organic pollutants and consequently to find other uses for such potential photocalatysts semiconductor nanomaterials.

References

[1] Deraz NM (2009) Influence of ceria on physicochemical, surface and catalytic properties of alumina supported manganese catalysts. Colloids and Surfaces A: Physicochemical and Engineering Aspects 335:8-15

[2] Gorte RJ (2010) Ceria in Catalysis: From Automotive Applications to the Water–Gas Shift Reaction. Reactors, Kinetics, And Catalysis 56:1126-1135

[3] Petar D, Levec J, Pintar A (2008) Effect of structural and acidity/basicity changes of CuO–CeO_2 catalysts on their activity for water–gas shift reaction. Catalysis Today 138:222–227

[4] Perrichon V, Laachir A, Bergeret G, Frety R, Tournayan L (1994) Reduction of cerias with different textures by hydrogen and their reoxidation by oxygen. J Chem Soc Faraday Trans 90:773-781

[5] Ricken M, Nolting J, Riess I (1984) Specific heat and phase diagram of nonstoichiometric ceria (CeO_{2-x}). J Solid State Chem 54: 89-99

[6] Korner R, Ricken M, Nolting J, Riess I (1989) Phase transformations in reduced ceria: Determination by thermal expansion measurements. J Solid State Chem 78:136-147

[7] Shi C, Yang L, Cai J (2007) Cerium promoted Pd/HZSM-5 catalyst for methane combustion. Fuel 86:106-112

[8] Trovarelli A, Leitenburg C, Boaro M, Dolcetti G (1999) The utilization of ceria in industrial catalysis. Catalysis Today 50:353-367

[9] Trovarelli A (2002) Structural properties and nonstoichiometric behavior of CeO_2. In: Catalysis by Ceria and Related Materials, Imperial College Press

[10] Ghandi HS, Piken AG, Shelef M, Deloch RG (1976) Automotive Engineering Congress and Exposition, Detroit, MI. SAE Paper 760201.

[11] Su EC, Montreuil CN, Rothschild WG (1985) Oxygen storage capacity of monolith three-way catalysts. Appl Catal 17:75-86

[12] Su EC, Rothschild WG (1986) Dynamic behavior of three-way catalysts. J Catal 99:506-510

[13] Engler B, Koberstein E, Schubert P (1989) Automotive exhaust gas catalysts: Surface structure and activity. Appl Catal 48:71-92

[14] Kacimi S, Barbier J Jr, Taha R, Duprez D (1993) Oxygen storage capacity of promoted Rh/CeC_2 catalysts. Exceptional behavior of RhCu/CeO_2. Catal Lett 22:343-350

[15] Duprez D, Descorme C, Birchem T, Rohart E (2001) Oxygen Storage and Mobility on Model Three-Way Catalysts. Topics in Catalysis 16-17: 49-56

[16] Rao R, Mishra BG (2003) Structural, redox and catalytic chemistry of ceria based materials. Bulletin of the Catalysis Society of India 2:122-134

[17] The Southland's War on Smog: Fifty Years of Progress Toward Clean Air (1997) http://www.aqmd.gov/news1/Archives/History/marchcov.html. Acessed 20 July 2010.

[18] Houdry EJ, Ardmore P (1952) Catalytic structure and composition. US patent 2742437.
[19] Christian Böhme (2007) The Success of the Automotive Catalytic Converter. http://www.catalysts.basf.com. Acessed 20 July 2010.
[20] Ono M, Okumura A (2010) Exhaust gas putifying catalyst and methos for purifying exhaust gas using catalyst. Patent WO 2010/044454.
[21] Satou A, Takeuchi M, Hiraku K et al Exhaust gas purification catalyst. Patent WO 2010/001215 A2.
[22] Miura M Exhaust gas purifying catalyst and process for producing it. Patent WO 2007/007889 A1.
[23] Kikuchi H, Nakamura M, Wakamatsu H et al Exhaust gas purification catalyst and method for production thereof. Patent WO 2006/054404 A1.
[24] Blake DM, Maness PC, Huang Z, Wolfrum EJ, Huang J (1999) Application of the Photocatalytic Chemistry of Titanium Dioxide to Disinfection and the Killing of Cancer Cells. Separation and Purification Methods 28:1-50
[25] Doshi J, Reneker DH (1995) Electrospinning process and applications of electrospun fibers. Journal of Electrostatics 35:151-160
[26] Shin YM, Hohman MM, Brenner MP, Rutledge GC (2001) Experimental characterization of electrospinning: the electrically forced jet and instabilities. Polymer 42:09955-09967
[27] Hashimoto K, Irie H, Fujishima A (2005) TiO_2 Photocatalysis: A Historical Overview and Future Prospects. Japanese Journal of Applied Physics 44: 8269–8285
[28] Photocatalysis: An Analysis of Its Applications and Market Potential (1998) EPRI, Palo Alto, CA. TR-111898
[29] Gueneau L, Rondet M, Besson S et al. Substrate with a self-cleaning coating. Patent WO 2003/087002.
[30] Funakoshi K, Nonami T (2007) Photocatalytic treatments on dental mirror surfaces using hydrolysis of titanium alkoxide. Journal of Coatings Technology and Research 4: 327-333
[31] Chung CJ, Lin H, Tsou HK, Shi ZY, He JL (2007) An antimicrobial TiO_2 coating for reducing hospital-acquired infection. Journal of Biomedical Materials Research Part B: Applied Biomaterials 85B(1): 220-224
[32] Veronovski N, Rudolf A, Smole MS, Kreže T, Geršak J (2009) Self-cleaning and handle properties of TiO_2-modified textiles. Fibers and Polymers 10:551-556
[33] Shen G, Chen PC, Ryu K, Zhou C (2009) Devices and chemical sensing applications of metal oxide nanowires. J Mater Chem 19:828-839
[34] Haggfeldt A, Bjorksten U, Lindquist SE (1992) Photoelectrochemical studies of colloidal TiO_2-films: the charge separation process studied by means of action spectra in the UV region. Sol Energy Mater Sol Cells 27:293-304
[35] Hermass JM (1999) Heterogeneous photocatalysis: fundamentals and applications to the removal of various types of aqueous pollutants. Catal Today 53:115-129
[36] Ha HK, Yosimoto M, Koinuma H, Moon B, Ishiwara H (1996) Open air plasma chemical vapor deposition of highly dielectric amorphous TiO_2 films. Appl Phys Lett 68:2965-2967
[37] Bahtat A, Bouderbala M, Bahtat M, Bouazaoui M, Mugnier J, Druetta M (1998) Structural characterisation of Er^{3+} doped sol–gel TiO_2 planar optical waveguides. Thin Solid Films 323:59-62.
[38] Desu SB (1992) Ultra-thin TiO_2 films by a novel method. Mater. Sci. Eng. B 13: 299-303

[39] Alves AK, Berutti FA, Clemens FJ, Graule T, Bergmann CP (2009) Photocatalytic activity of titania fibers obtained by electrospinning. Materials Research Bulletin 44:312-317
[40] Porter F (1991) Zinc Handbook: Properties, Processing, and Use in Design. Marcel Dekker Inc, New York.
[41] International Zinc Association. http://www.zincworld.org. Accessed 02 August 2010
[42] Arnold MS, Avouris P, Pan ZW et al (2003) Field-effect transistors based on single semiconducting oxide nanobelts. J Phys Chem B 107:659-663
[43] Keem K, Kim H, Kim GT et al (2004) Photocurrent in ZnO nanowires grown from Au electrodes. Appl Phys Lett 84: 4376-4379
[44] Lee CJ, Lee TJ, Lyu S C et al (2002) Field emission from well-aligned zinc oxide nanowires grown at low temperature. Appl Phys Lett 81:3648-3650
[45] Huang MH, Mao S, Feick H, Yan HQ et al (2001) Room temperature ultraviolet nanowire nanolasers. Science 292:1897-1899
[46] Park WI, Jun YH, Jung SW et al (2003) Excitonic emissions observed in ZnO single crystal nanorods. Appl Phys Lett 82:964-967
[47] Park JH, Choi HJ, Choi YJ et al (2004) Ultrawide ZnO nanosheets. J Mater Chem 14:35-36
[48] Wang ZL (2004) Nanostructures of zinc oxide. Mater Today 7:26-33
[49] Park JH, Choi HJ, Park JG (2004) Scaffolding and filling process: a new type of 2D crystal growth. J Cryst Growth 263:237-242
[50] European Commission (2006) European survey on success factors, barriers and needs for the industrial uptake of nanomaterials in SMEs. http://www.nanoroad.net/download/sme_survey.pdf. Accessed 02 August 2010.
[51] J Lacson (2000) Inorganic zinc chemicals. Chemical Economics Handbook, electronic release. SRI International, Menlo Park CA, USA.
[52] FDA (2006) Public meeting on nanotechnology materials in FDA regulated products, Dept. of Health and Human Services.
[53] P Harrison (2007) Emerging challenges: nanotechnology and the environment. GEO Year Book. United Nations Environment Programme (UNEP), Phoenix Desing Aid, Denmark
[54] Lux Research (2004) Revenue from nanotechnology-enabled products to equal IT and telecom by 2014. Press release
[55] Inframat® Advanced MaterialsTM LLC. http://www.advancedmaterials.us/30N-0801.htm. Accessed 02 August 2010.
[56] Nanophase Technologies Corporation. http://www.nanophase.com/products/details.aspx?ProductId=7. Accessed 02 August 2010.
[57] Evonik Industries. http://nano.evonik.com. Accessed 02 August 2010.
[58] NaBond Technologies Co. http://www.nabond.com/ZnO_nanopowder.html. Accessed 02 August 2010.
[59] M K Impex Canada . http://www.mknano.com. Acessed 02 August 2010.
[60] Nano-Infinity Nanotech Co., Ltd. http://www.b2bnano.com/nanomaterial.htm. Accessed 02 August 2010.
[61] Nanostructured & Amorphous Materials Inc. http://www.nanoamor.com/inc/sdetail/18824. Accessed 02 August 2010.
[62] Hoffmann MR, Martin ST, Choi W et al (1995) Environmental Applications of Semiconductor Photocatalysis. Chem Rev 95:69-96

[63] Li D, Haneda H, (2003) Morphologies of zinc oxide particles and their effects on photocatalysis. Chemosphere 51:129-137
[64] Dindar B, Icli S (2001) Unusual photoreactivity of zinc oxide irradiated by concentrated sunlight. J Photochem Photobiol A: Chem. 140:263-268.
[65] Ahmed S, MG Rasula, Wayde NM et al (2010) Heterogeneous photocatalytic degradation of phenols in wastewater: A review on current status and developments. Desalination 261:3-18
[66] Parida KM, Parija S (2006) Photocatalytic degradation of phenol under solar radiation using microwave irradiated zinc oxide. Solar Energy 80:1048-1054
[67] Hariharan C (2006) Photocatalytic degradation of organic contaminants in water by ZnO nanoparticles: Revisited. Applied Catal A:General 304:55-61
[68] Dijken V, Janssen AH, Smitsmans MHP et al (1998) Size-Selective Photoetching of Nanocrystalline Semiconductor Particles. Chem Mater 10:3513-3522
[69] Neppolian S, Sakthivel B, Arabindoo M et al (1998) Photocatalytic Degradation of Textile Dye Commonly Used in Cotton Fabrics. Stud Surf Sci Catal 113:329–335
[70] Gouvea CAK, Wypych F, Moraes SG et al (2000) Semiconductor-assisted photocatalytic degradation of reactive dyes in aqueous solution. Chemosphere 40:433-440.
[71] Dindar S, Icli J (2001) Unusual photoreactivity of zinc oxide irradiated by concentrated sunlight. Photochem Photobiol A: Chem 140:263-268
[72] Behnajady MA, Modirshahla N, Hamzavi R (2006) Kinetic study on photocatalytic degra-dation of C.I. Acid Yellow 23 by ZnO photocatalyst. J Hazard Mater B 133:226-232.
[73] Qiu R, Zhang D, Mo Y et al (2008) Photocatalytic activity of polymer modified ZnO under visible light irradiation. J Hazard Mater 156:80-85
[74] Trommer RM, Alves AK, Bergmann CP (2010) Synthesis, characterization and photocata-lytic property of flame sprayed zinc oxide nanoparticles. J Alloys Compounds 491:296-300
[75] Roth P (2007) Particle synthesis in flames. Proc. Comb. Inst. 31:1773-1788
[76] Strobel R, Baiker A, Pratsinis SE (2006) Aerosol flame synthesis of catalysts Adv. Powder Technol 17:457-480
[77] Wang ZL;(2004) Zinc oxide nanostructures: growth, properties and applications. J. Phys Condens Matter 16:829-858
[78] Wang N; Cai Y.; Zhang RQ (2008) Growth of nanowires. Mater Sci and Engineering 60:1-51
[79] Wang ZL (2004) Nanostructures of zinc oxide. Mater Today 7:26-33
[80] Polarz S, Pueyo CL, Krumm M (2010) The molecular path to inorganic materials - zinc oxide and beyond. Inorganica Chimica Acta. in press.
[81] Meulenkamp EA (1998) Synthesis and Growth of ZnO Nanoparticles. J Phys Chem B 102:5566-5572
[82] Dodd A, McKinley A, Saunders M et al (2006) A comparative evaluation of the photocatalytic and optical properties of nanoparticulate ZnO synthesised by mechanochemical processing. J. Nanopart Res 10:243-248
[83] Beydoun D, Amal R, Low G et al (1999) Role of Nanopaiticles in Photocatalysis. J. Nanopart. Res. 1:439-458
[84] Dodd A, McKinley A, Tsuzuki T et al (2009) Tailoring the photocatalytic activity of nanoparticulate zinc oxide by transition metal oxide doping. Mater Chem and Phys 114:382–386

[85] Ricci A, Chrétien M, Maretti L et al (2003) TiO_2 promoted mineralization of organic sunscreen in water suspension and sodium dodecyl sulfate micelles. Photochem Photobiol Sci 2:487–492

[86] Wang Y, Li X, Wang N et al (2008) Controllable synthesis of ZnO nanoflowers and their morphology-dependent photocatalytic activities. Separation and Purification Technology 62:727-732

Abbreviations

A/F - Air-Fuel ratio
BET - Brunauer–Emmett–Telle
FCC - Face Centered Cubic
FESEM – Field Emission Scanning Electron Microscopy
ZnO-FSb - Flame Synthesis based nano-ZnO
HPO - Heterogeneous Photocatalytic Oxidation
MO - Methyl Orange
MB - Methylene Blue
OSC - Oxygen Storage Capacity
PMMA - Polymethylmetacrilate
SEM – Scanning Electron Microscopy
TEM – Transmission Electron Microscopy
ZnO TEb - Thermal Evaporation based nano-ZnO
TWC - Three Way Catalyst ()
UV - Ultraviolet

8 Nanoreinforcements for Nanocomposite Materials

Sérgio Henrique Pezzin[1], Sandro Campos Amico[2], Luiz Antônio Ferreira Coelho[1], and Mônica Jung de Andrade[2]

[1] Centro de Ciências Tecnológicas, Universidade do Estado de Santa Catarina, 89219710, Joinville, Brazil
[2] Departamento de Engenharia de Materiais, Universidade Federal do Rio Grande do Sul, 91501970, Porto Alegre, Brazil
E-mail: pezzin@joinville.udesc.br

Abstract. The range of materials currently available as fillers for nanocomposites include nanoparticles, nanoplatelets, carbon nanotubes, nanofibers, and, more recently, graphenes or even a combination of these unique materials. Several companies already supply a variety of ceramic, metal and polymer nanocomposite products that reach into many industrial sectors, such as aerospace, energy and sporting goods. Polymer nanocomposites attract most of the research and development effort and, as a result, more traditional microreinforced materials are already being substituted. Indeed, the combination of nanomaterials with thermoplastic or thermoset polymers may lead to very interesting mechanical or physical properties. As shown in this chapter, these nanofillers are slowly finding their way into mainstream commercial use and their even wider application can be foreseen when difficulties regarding cost, homogeneous dispersion and poor adhesion to the host matrix are further minimized.

Keywords: nanofillers, nanoreinforcements, polymers, industrial applications.

8.1 Introduction

Nanocomposites, broadly defined as nanofillers bonded to a matrix, have been reported for years but are recently gaining momentum in mainstream commercial use. Research focusing on nanotechnology has allowed several companies to produce ceramic, metal and polymer nanocomposites with outstanding properties in comparison with microstructured composites. The benefits encompass improved mechanical properties, scratch resistance, barrier properties, fire resistance, dimensional stability and, in some cases, faster processing. Besides, a small amount of nanofillers can cut weight and cost compared with the usual loading of conventional fillers.

Carbon-based nanomaterials as vapor-grown carbon fiber (VGCF) [1], Carbon NanoSphere Chain™ [2], fullerenes [3] and carbon nanotubes (CNTs) [3,4,5,6] have been industrialized as fillers or nanoreinforcements due to their low density, high strength, high electrical and thermal conductivities. More recently, graphene has also been investigated as nanoreinforcement [7,8,9,10].

Highly thermally conductive nanocomposites (carbon, silicon carbide, copper, aluminum or epoxy matrices) reinforced by Pyrograf®-I VGCF are being commercialized by Nanographite Materials [1], a joint venture between Applied Sciences, Inc., (in USA) and GSI Creos Corporation (in Japan). The reported thermal conductivity varied from 310 W m^{-1} K^{-1} (20 vol.% of Pyrograf®-I VGCF in SiC matrix) to 910 W m^{-1} K^{-1} (70 vol.% of Pyrograf®-I VGCF in carbon matrix) [1], enabling thermal management in aerospace applications, for example.

Nanocomposites with Carbon NanoSphere Chain™ (CNSC) [2] have been investigated for applications such as super capacitors, protective armor, radar cross-sectional absorption and electrical anti-icing. Fullerenes [3] can be used as reinforcement in nanocomposites with applications that can vary from solar cells [11] to proton exchange membranes of fuel cells [12] or electrocatalytic [13] applications. Nanoscale graphene platelets (NGP), which are composed of one or more layers of graphene planes with platelet thickness from 0.34 to 100 nm [14], have been used for nanocomposite applications including solar cells [7], biosensors [8], supercapacitors [9] and electrodes for lithium ion batteries [10].

CNTs are among the most popular nanofillers. The number of publications focusing on CNTs grows exponentially and more than 60000 publications in the last 19 years may be found in the ISI (Science Citation Index) database [15]. CNTs are considered as nearly one-dimensional structures due to their very high aspect ratio (up to 10^3-10^4), i.e. the ratio between length (usually in the order of micrometers [16]) and diameter (usually between 1 and 20 nm [16]). CNTs can be composed of one, two or many concentric tubes, known as single-walled (SWCNTs), double-walled (DWCNTs) and multi-walled (MWCNTs) CNTs, respectively [16]. Their unique properties allow the production of nanocomposites with applications in the energy and (opto-) electronic fields, and also as heating or structural elements [16].

Not surprisingly, this new generation of composites is already being produced and commercialized by companies such as Nanocyl [4], Nanotech [5] and Nanolab [6]. Nanocyl's CNTs composite technology [4] focuses on polymer-based products, such as EPOCYL™ Epoxy Resins, SIZICYL™ Sizing Agent and PREGCYL™ Pre-preg Materials, whereas the Korean Nanotech [5] manufactures WC-Co nanocomposite powder (through a spray conversion method) and sinters nanocomposites with excellent hardness, toughness and wear resistance. The latter [6] developed hot pressing and pressureless sintering protocols to produce a high toughness (fracture toughness of ~5 MPa $m^{-1/2}$ at 4 wt.% CNTs) boron carbide composite containing CNTs (B_4C-CNT composite).

The global consumption of nanocomposites was estimated in 64,570 metric tons with a value of nearly $460 million in 2009. By 2014, the market should exceed 214,000 metric tons and $1.38 billion, a compound annual growth rate of ca. 27%. The major part of this market corresponds to clay-based nanocomposites with a consumption of approximately 44,930 metric tons and $227 million in

2009, which is expected to increase to 181,000 metric tons and $692 million by 2014, a compound annual growth rate of c.a. 32% The overall world market for ceramic-containing nanocomposites was approximately 10,662 metric tons with a value of $49.8 million in 2008 and is expected to grow to 17,388 metric tons and $145.2 million in 2014, a compound annual growth rate of 12.5% [17]. A large part of this market and most research works found in the literature are devoted to polymer-matrix nanocomposites, on which this chapter will focus.

8.2 Nanoreinforcements for Polymer Composites

The most widely employed composites nowadays are those with micron-sized particles/fibers as reinforcement. It should be emphasized that, in these composites, a large filler content is required and this can adversely affect some properties of the neat matrix. In fact, traditional microcomposites may already have reached their limit regarding the optimization of properties, e.g. in some cases, stiffness and/or strength are achieved at the expense of fracture toughness.

About twenty years ago, researchers showed that novel materials could be produced with tailored properties using nanofillers. In this scenario, two questions were raised: (i) What is the difference between nanofillers and microsized fillers? (ii) How do microcomposites compare to nanocomposites? It should be noted that smaller sizes combined with homogeneous dispersion decrease the likelihood of finding large stress concentration sites within the material. In addition, nanofillers can provide exceptionally large interfacial area in composites. Wu *et al.* [18] pointed out that an interphase 1-nm thick represents roughly 0.3% of the volume of polymer in microcomposites, whereas it can reach 30% of the total volume in nanocomposites. Indeed, this huge interfacial area can reach 1000 m^2/g filler [11,18,19,20,21,22]. Because of that and the unique properties of nanoreinforcements, the incorporation of a small amount of CNTs in polymers, generally between 0.1 and 5 wt.%, can potentially yield structural materials with exceptionally high stiffness and strength [23,24].

Polymer-matrix nanocomposites can be classified as structural or functional depending on the role of the nanofiller in each situation. For structural composites, the mechanical properties of the nanoparticle, such as high Young's modulus, tensile strength and elongation at break, and the ability to resist compression and distortion can be used to produce lightweight structural materials. On the other hand, in functional composites, other interesting properties are exploited, such as high electrical and thermal conductivity, required in the development of thermal resistant materials, sensors, electrical conductors, photoemitters, electromagnetic shields or energy accumulators [25,26].

8.2.1 Nanoreinforcements in Thermoplastic Polymers

The main goal when reinforcing thermoplastics is usually to develop a material which combines high toughness and large plastic elongation to break with high stiffness and strength. The general types of nanofillers used in these nanocomposites may be classified into layered, particulated and fibrous materials [20], which are briefly detailed below.

(i) Nanoplatelet-reinforced composites

The most common nanofillers in this class are layered silicates. A layer of silicate clay is about 1-nm thick with platelets of around 100–200 nm in width. According to Yu *et al.* [27], clays have been extensively studied because they are naturally occurring minerals that are readily commercially available, exhibit a layered morphology with high aspect ratio and large specific surface area, and also due to their significant cation exchange capacity. More recently, attention has been drawn to the exfoliation of graphite sheets to obtain graphene which can also be used to reinforce polymeric materials [21,22].

Perhaps the first successful thermoplastic nanocomposite was a polyamide 6 (nylon® 6 or PA 6)/clay hybrid developed at Toyota Central Research & Development Co. Inc. in 1994 and used as a timing belt cover for the Toyota Camry automobile. This composite with 4.2 wt.% clay had twice the modulus, 50% higher strength and 80 °C higher heat distortion temperature compared to neat PA 6. It should be pointed out that they had been working on this subject since the beginning of the previous decade [28]. After that, polymer/clay nanocomposites attracted much interest from the academic and industrial sectors worldwide.

It is common sense that exfoliated clays yield better mechanical properties such as elastic modulus and tensile strength than intercalated ones. Fornes *et al.* [29] showed that when higher molecular weight PA 6 was used, the elastic modulus of its nanocomposite with montmorrilonite (MMT) increased, even though the stiffness of the neat matrix remained nearly constant. A trend towards an increase in tensile strength of the nanocomposites was also observed and the authors argued that a high molecular weight matrix is more effective in exfoliating clays. Nevertheless, some reports in the literature indicate that intercalated clays are able to yield higher fracture toughness than exfoliated structures [27].

Other thermoplastics have also been investigated. For instance, neat polymethylmethacrylate (PMMA) showed an increase in Young's modulus from 3.0 to 4.6 GPa when 10 wt.% intercalated sodium modified montmorillonite clays (MMT-Na^+) was added. For neat polystyrene (PS), the respective increase was from 1.2 to 1.9 GPa. Regarding tensile strength, the increase for PMMA was from 53.9 to 62.0 MPa, whereas for PS a decrease from 28.7 to 23.4 MPa was reported [30,31].

(ii) Nanoparticle-reinforced composites

Most studies in this class focus on semi-crystalline polymers and inorganic nanoparticles, usually showing an embrittling effect. Different particles have been employed to prepare nanocomposites, such as: metallic (Al, Fe, Au, and Ag), metal oxides (ZnO, Al_2O_3. $CaCO_3$, and TiO_2) and nonmetal oxide (SiO_2). There are companies such as Degussa and Nanophase which supply nanoparticles of various sizes for different applications. In addition, there are numerous surface treatments available, either due to physical interaction or due to chemical reaction [32,33].

Kausch and Michler [34] investigated the effect of the size and size-distribution of the nanoparticle on the mechanical behaviour of nanocomposites produced with several polymers (PS, PMMA and polycarbonate - PC) and alumina or SiO_2. In PMMA/SiO_2 nanocomposites, K_{Ic} increased from 30 for neat PMMA to 67 MPa.mm$^{1/2}$ for a composite with 5 vol.% SiO_2.

(iii) Nanofiber-reinforced composites and Carbon nanotube-reinforced composites

Frogley *et al.* [35] doped a silicone-based elastomer with both SWNT and carbon nanofibre. An increase of 200% and 120% in elastic modulus was achieved for SWCNT and nanofibres, respectively, in comparison to the neat material. Coleman *et al.* [36] reported that when polyvinyl alcohol (PVA) crystallizes around carbon nanotubes there is an increase in interfacial stress transfer. For PP, functionalization of the interface was carried out by adding covalently chlorinated polypropylene chains to the nanotubes. For PVA, increase of 370% was observed in Young´s modulus, 430% in tensile strength and 170% in toughness, whereas for PP the increase reached 310%, 390% and 440%, respectively, using less than 1 wt.% of carbon nanotubes in both cases.

8.2.2 Nanoreinforcements in Thermoset Polymers

Many different nanoparticles have been used to reinforce thermoset polymers, such as alumina, silica, whiskers and nanocarbon black. However, this section will focus on nanoclays, carbon nanotubes and graphene, which are amongst the most researched nanoreinforcements in the last few years.

Regarding thermoset polymers, epoxy resins are the common choice for nanocomposites used for high performance applications in structures, coatings and adhesives. Highly crosslinked epoxies exhibit high modulus, excellent resistance to solvents and to corrosion, good adhesion, dimensional stability, good performance at high temperatures, relatively low cost and easy processing. On the other hand, these resins are quite brittle after curing [37]. A large variety of epoxy resins are available in the market with final characteristics that depend on the particular epoxy monomer, the curing agent, the additives in the composition and the curing process itself. Many research groups worldwide have been studying epoxy-matrix composites reinforced with nanoparticles [26,32,38]. Nevertheless, it is important to add that unsaturated polyester may become the primary choice of thermoset resin for nanocomposites in the future if these nanofillers are increasingly used to enhance or replace glass fibers as reinforcement in a wider range of less expensive applications.

(i) Nanoplatelet-reinforced composites

This class of nanocomposite is well established and a number of commercial products are already available. Polymer/nanoclay materials can be broadly divided into three categories: conventional (flocculated), intercalated and exfoliated, the latter usually showing better properties [39,40,41]. The combination of organic functionalities from the polymers with inorganic materials, especially clays, can yield several singular properties to the composites, such as flexibility, durability, toughness, as well as wear resistance and high thermal stability and hardness [21,40,42,43].

The most important group of layered clay minerals for nanocomposites is smectites, negatively charged silicate layers and exchangeable interlayer cations that are very convenient as host structures for intercalation. Montmorillonites and hectorites are two typical examples of this clay group [22]. Sodium modified montmorillonite clays (MMT-Na^+) can be easily intercalated or exfoliated [44] and are worldwide supplied by a number of companies, including Southern Clay (Cloisite™ and Garamite™ lines), Nanocor (Nanomer™ line) and Elementis Specialties (Bentone and hectorite lines). One of their interesting applications is in sheet molding compound (SMC) of glass-reinforced thermosets, in which the addition of 0.5% of Cloisite® clay can reduce in more than 20% the composite density. There are many studies on the replacement of interlayer metal cations by organoammonium cations. This procedure turns the interlayer space strongly organophilic, producing the so-called organoclays [22], and can greatly enhance the interfacial adhesion between exfoliated organoclays and the polymer matrix.

In situ intercalative polymerization is a viable technique for the preparation of thermoset-based nanocomposites [22]. In such process, exfoliation of the organoclays is determined by several factors such as their nature, catalytic effect on the curing reaction and miscibility with the curing agent. There is a competition during curing between the resin that is intercalated in the clay and the bulk resin and, if polymerization occurs at similar rates in both processes, the curing heat is enough to overcome the attractive forces between silicate layers and, as a result, an exfoliated nanocomposite structure can be obtained.

Another layered reinforcement that has been the focus of various recent studies is graphene. The great interest in this material is perhaps a consequence of its expected large scale production [45,46]. A number of methods for the production of graphene have been reported in the literature, including the use of epitaxial growth and chemical vapour deposition (CVD) [47,48]. A low-cost and potentially upscalable method comprises the production of functionalized graphene sheets (FGS) in relatively large amount via thermal expansion of graphite oxide (GO), which is produced by the oxidation of graphite with strong oxidizers or from intercalation compounds [48].

Some work has been done on the preparation of epoxy/graphene composites, usually comprising the production of films by casting [27,49]. Dispersion is probably the most fundamental issue in the development of epoxy/graphene nanocomposites. The exfoliation of graphite is an alternative route for the separation of graphene layers [46], but the lack of functional groups in graphene sheets leads to reaggregation in micrometric "clusters".

A great effort has been devoted to the dispersion of these nanometric particles in polymer matrices using a variety of techniques [50]. A common method to disperse graphene sheets (GS) in epoxy resins is via sonication. High-power ultrasound pulses can separate graphene sheets and disperse them in the uncured resin, yielding GS metastable suspensions [51]. The optimal sonication conditions (i.e. time and power) are yet to be determined and will certainly vary according to concentration and initial graphene exfoliation, as well as the presence of solvents and surfactants.

The use of FGS, a strategy to increase interfacial adhesion, represents an important research field in nanocomposites reinforced with graphene. Functionalization can also assist the dispersion of graphene sheets in polymer matrices, considering that the graphene layers can be separated during the process. Moreover, the introduction of "tailor-made" functional groups (e.g. amino-, carboxyl- or glycidil-) may lead to covalent bonds between graphene and the resin, enhancing the stress transfer at the interface [51].

(ii) Carbon nanotube-reinforced composites

Carbon nanotubes (CNTs) have been extensively investigated as fillers for polymer nanocomposites since their "discovery" in 1991 [52] and a number of reviews can be found on this subject [23,24,26,38,53]. Obtaining nanocomposites with superior properties depends on the CNT dispersion (without compromising their integrity) and the nanotube/polymer matrix adhesion at the interface [54,55]. Companies like the north-american NanoSperse [56] are developing technologies for the uniform dispersion of nanomaterials used for advanced thermoset composites.

Despite the enormous research effort already carried out, the expected significant mechanical reinforcement of polymer matrices by CNTs is still to be achieved. It is well-known that the addition of a small amount of CNTs in highly crystalline thermoplastic polymers can increase the Young's modulus in more than 100% [23]. For thermosets, however, the scenario is quite controversial. The most significant enhancement in Young's modulus of epoxy matrices appears to have been 88%, using 0.25 wt.% of carboxylated SWCNTs [23,57]. Nevertheless, it must be pointed out that CNTs do increase the electrical conductivity of thermoset polymers. For epoxy resins, for instance, an enhancement of $10^4 – 10^5$ S. cm^{-1} can be achieved. There is also evidence that MWCNTs may yield percolating networks, being therefore suitable for applications requiring higher electrical conductivity [58]. This makes epoxy/CNT nanocomposites strong candidates for electrostatic discharge (ESD) materials, which are used in various fields, including the aeronautic sector.

(iii) Other nanoreinforced composites

Nanocomposites combining nanoclays and carbon nanotubes in polymer matrices have been recently studied. These materials can be divided into two categories: i) carbon nanotubes synthesized (usually by chemical vapor deposition) on clay surfaces and later incorporated into a polymer matrix [59,60]; ii) clays and carbon nanotubes separately introduced in the same polymer matrix [40,42,44]. Lan and Lin [61] reported that the mixture of MMT-Na^+ and CNTs enhances the dispersion of nanotubes and concluded that clay dimensions and ionic interactions between clays and CNTs are the most important factors to influence dispersion. An increase of 100% in impact strength and 70% in Vickers hardness was recently reported for epoxy/clays/CNTs nanocomposites [60], and Zhang and Wang [44] found 50% enhancement in Young's modulus and better clay exfoliation in the presence of CNTs.

8.2.3 Examples of Commercial Applications

A variety of nanocomposites using clays are currently available. For instance, Nylon nanocomposites produced by Bayer, Honeywell Polymer and Toyota Company have been applied to timing belt cover and engine cover in cars. Polyolefin nanocomposite applications can also be found in products of General Motors and Southern Clay Products. The latter produces Cloisite, which is an organophilic clay that lead to 23% increase in tensile strength, 69% in tensile modulus and 56% in flexural modulus for a loading of 5 wt.% in PA 6 [62]. Sud-Chemie developed Nanofil, which was designed to improve mechanical and thermal properties of thermoplastics. For example, tensile modulus is improved by 25% and tensile strength by 14% when 2% of Nanofil® is added to PA 6 [63].

Bentone, produced by Elementis Specialties, is employed to improve the mechanical properties of thermosets and thermoplastics (e.g. polyester, polystyrene and polyamides) [64]. A product developed by Noble Polymer, called Forte nanocomposite, with improved stiffness and impact properties, was applied to the seat back of cars [65].

Although great commercial interest in polymer nanocomposites focuses on thermoplastic, there are several thermoset nanocomposite products available on the market. Graphistrength® [66], developed by Zyvex Performance Materials and Arkema, is a liquid epoxy dispersion designed to enhance the mechanical strength of epoxy composites. These epoxy resin masterbatches (based on either diglycidyl ether of bisphenol-A, DGEBA, or diglycidyl ether of bisphenol-F, DGEBF) with 3.0 wt.% MWCNT can be used to produce high strength composites, coatings and adhesives for wind turbine blades, tennis rackets, baseball bats, skis and bicycle frames.

Nanocor® [67] supplies Nanomer® Nanoclays, which are microfine powders that can be directly incorporated into epoxy or urethane resin systems, and Aardvark Polymers supplies epoxy masterbatches with a number of nanofillers, including carbon nanotubes (both SWCNT and MWCNT) and nanoclays. Hyperion [68] offers a series of MWCNTs masterbatches (Fibril®), typically 15-20 wt.%, dispersed into polymers, including DGEBA epoxy. NanoLok™ PT, commercialized by InMat Inc. [69], is an environmentally friendly aqueous suspension of montmorillonite clay in sulfonated polyester resin that may be used as a gas barrier coating when applied to polymer films.

The University of Cincinnati developed a multifunctional hydrophobic electrically conductive and magnetic CNSC-polymer (epoxy or polyurethane) nanoskin that can be used to produce anti-icing heating film, low impedance piezoresistivity sensors, erosion resistant coatings, electronic storage devices and to build artificial organs [70].

Nanocyl's CNTs composite technology [4] has marketed a variety of polymer-based nanocomposites, including the products named BIOCYL™ and

THERMOCYL™ (barriers against fouling and flame, respectively) and PREGCYL™ Pre-preg Materials. The NanoComposites Inc. company [71] is also working on potential applications of composites with CNTs to increase mechanical strength, abrasion resistance, elongation and fatigue time, and also to enhance pressure maintenance and thermal stability. A new generation of CNT-polymer nanocomposite textiles with electrical and thermal conductivity was patented by Australia's national science agency - CSIRO [72], in collaboration with the NanoTech Institute of the University of Texas, who developed the technology of CNT yarns [73].

Finally, Vorbeck Materials Corp. [74] has recently launched graphene dispersions in a wide range of thermoplastics, rubbers and thermosets. These composites are intended to have superior mechanical and electrical properties in comparison with CNT nanocomposites, along with barrier and thermal properties similar to nanoclay composites.

8.3 Concluding Remarks

The examples shown here illustrate the multiplicity of applications that are already a reality for nanocomposites. However, it has to be pointed out that microcomposites are still, by far, the primary materials for structural applications for high mechanical performance. Nevertheless, many applications which rely on the physical or even mechanical characteristics of the part are already benefiting from the unique behavior that a given material can achieve at a significantly low content of nanoreinforcement.

The nanofillers available, including nanoparticles, nanoplatelets, carbon nanotubes, nanofibers, and more recently graphenes or even a combination of these materials, are indeed finding their way into mainstream commercial use. Several companies are already supplying a variety of ceramic, metal and mainly polymer nanocomposite products that reach into many industrial sectors, from solar cells, fuel cells or electrocatalytic applications to super capacitors, biosensors and (opto-)electronics, to name a few.

References

[1] http://www.apsci.com/ngm-about.html Accessed June 2010
[2] http://www.cleantechnano.com/applications.html Accessed June 2010
[3] http://www.mercorp.com/products2.htm Accessed June 2010
[4] http://www.nanocyl.com/ Accessed May 2010
[5] http://www.nanopowder.co.kr/en/ Accessed May 2010
[6] http://www.nano-lab.com/carbonnanotubeproducts.html Accessed May 2010
[7] Tang YB, Lee CS, Xu J, Liu ZT, Chen ZH, He ZB, Cao YL, Yuan GD, Song HS, Chen LM, Luo LB, Cheng HM, Zhang WJ, Bello I, Lee ST (2010) Incorporation of Graphenes in Nanostructured TiO_2 Films via Molecular Grafting for Dye-Sensitized Solar Cell Application. ACS Nano 4:3482-3488

[8] Hong WJ, Bai H, Xu YX, Yao ZY, Gu ZZ, Shi GQ (2010) Preparation of Gold Nanoparticle/Graphene Composites with Controlled Weight Contents and Their Application in Biosensors. Journal of Physical Chemistry C 114:1822-1826

[9] Zhang YP, Li HB, Pan LK, Lu T, Sun Z (2009) Capacitive behavior of graphene-ZnO composite film for supercapacitors. Journal of Electroanalytical Chemistry 634:68-71

[10] Lee SW, Yabuuchi N, Gallant BM, Chen S, Kim B-S, Hammond PT, Shao-Horn Y (2010) High-power lithium batteries from functionalized carbon nanotube electrodes. Nature Nanotechnology 5:531-537

[11] Yu J, Zhang ZH, Ni Y, Lu YB Xiong YQ, Xu WJ (2007) Preparation and characterization of a novel composite based on hyperbranched polysilane and fullerene. Journal of Applied Polymer Science 105:821-826

[12] Wang HB, De Sousa R, Gasa J, Tasaki K, Stucky G, Jousselme B, Wudl F (2007) Fabrication of new fullerene composite membranes and their application in proton exchange membrane fuel cells. Journal of Membrane Science 289:277-283

[13] Qi B, Yang XR (2008) Characterization and electrocatalytic application of a fullerene/ionic-liquid composite. Materials Letters 62:980-983

[14] Jang BZ, Zhamu A (2008) Processing of nanographene platelets (NGPs) and NGP nanocomposites: a review. J Mater Sci 43:5092-5101

[15] http://Apps.Isiknowledge.Com/Summary.Do?Qid=6&Product=Ua&Sid=3aeccodn9ekem18bama&Search_Mode=Generalsearch Accessed November 2010

[16] O'Connell MJ (2006) Carbon Nanotubes – Properties And Applications. Taylor &Francis, New York

[17] McWilliams A (2010) Nanotechnology: Nanocomposites, Nanoparticles, Nanoclays, and Nanotubes, BCC Research, Report Code: NAN021D http://www.bccresearch.comwww.bccresearch.com/report/NAN021D.html. Accessed August 2010

[18] Wu CL, Zhang MQ, Rong MZ, Friedrich K (2002) Tensile performance improvement of low nanoparticles filled-polypropylene composites. Composites Science and Technology 62:1327-1340

[19] Thostenson ET, Ren Z, Chou T-W (2001) Advances in the science and technology of carbon nanotubes and their composites: a review. Composites Science and Technology 61:1899-1912

[20] Thostenson ET, Li C, Chou T-W (2005) Nanocomposites in context. Composites Science and Technology 65:491-516

[21] Alexandre M, Dubois P (2000) Polymer-layered silicate nanocomposites: preparation, properties and uses of a new class of materials. Materials Science and Engineering R 28:1-63

[22] Pavlidou S, Papaspyrides CD (2008) A review on polymer–layered silicate nanocomposites. Progress in Polymer Science 33:1119-1198

[23] Coleman JN, Khan U, Blau WJ, Gun'ko YK (2006) Small but strong: A review of the mechanical properties of carbon nanotube–polymer composites. Carbon 44: 1624-1652

[24] Tjong SC (2006) Structural and mechanical properties of polymer nanocomposites. Materials Science and Engineering R 53:73-197

[25] Amico SC, Pezzin SH, Coelho LAF (2008) Nanocompósitos de matriz polimérica com nanotubos de carbono in Pohlmann AR, Petter CO, Balzaretti NM, Guterres SS (org.), Tópicos em Nanociência e Nanotecnologia, Editora da UFRGS

[26] Moniruzzaman M, Winey KI (2006) Polymer nanocomposites containing carbon nanotubes. Macromolecules 39:5194-5205

[27] Yu Z-Z, Dasari A and Mai Y-W (2007) Polymer-Clay Nanocomposites - A Review of Mechanical and Physical Properties In: Suresh G (ed) Advani Processing and Properties of Nanocomposites. World Scientific Publishing Co. Pte. Ltd, Singapure
[28] Masami Okamoto, (2004) Polymer/Clay Nanocomposites. In: H. S. Nawa (ed) Encyclopedia of Nanoscience and Nanotechnology, vol 8. H.S. Nalwa, New York
[29] Fornes TD, Yoon PJ, Keskkula H and Paul DR (2001) Nylon 6 nanocomposites: the effect of matrix molecular weight. Polymer 42:9929–9940
[30] Lee DC and Jang LW (1996) Preparation and characterization of PMMA–clay hybrid composite by emulsion polymerization. J Appl Polym Sci 61:1117-1122
[31] Noh MW and Lee DC (1999) Synthesis and characterization of PS–clay nanocomposite by emulsion polymerization. Polym Bull 42: 619-626
[32] Hussain F, Hojjati M, Okamoto M, Gorga RE (2006) Polymer-matrix Nanocomposites, Processing, Manufacturing, and Application: An Overview. Journal of Composite Materials 40:1511-1575
[33] Karger-Kocsis J and Zhang Z (2005) Structure-Property Relationships in Nanoparticle/Semicrystalline Thermoplastic Composites, Chapter 13, in Mechanical Properties of Polymers Based on Nanostructure and Morphology, edited by Michler GH, Baltá-Calleja FJ, Taylor & Francis
[34] Kausch HH, Michler GH (2007) Effect of nanoparticle size and size-distribution on mechanical behavior of filled amorphous thermoplastic polymers. Journal of Applied Polymer Science 105:2577-2587
[35] Frogley MD, Ravich D, Wagner HD (2003) Mechanical properties of carbon nanoparticle-reinforced elastomers. Composites Science and Technology 63:1647-1654
[36] Coleman JN, Cadek M, Blake R, Nicolosi V, Ryan KP, Belton C, Fonseca A, Nagy JB, Gunko YK and Blau WJ (2004) High-Performance Nanotube-Reinforced Plastics: Understanding the Mechanisc of Strength Increase. Advanced Functional Materials 14:791-798
[37] Thomas R, Durix S, Sinturel C, Omonov T, Goossens S, Groeninckx G, Moldenaers P, Thomas S (2007) Cure kinetics, morphology and miscibility of modified DGEBA-based epoxy resin - Effects of a liquid rubber inclusion. Polymer 48:1695-1710
[38] Chou T-W, Gao L, Thostenson ET, Zhang Z, Byun J-H (2010) An assessment of the science and technology of carbon nanotube-based fibers and composites. Composites Science and Technology 70:1-19
[39] Chin I, Thurn-Albrecht T, Kim H, Russel TP, Wang J (2001) On exfoliation of montmorillonite in epoxy. Polymer 42:5947-5952
[40] Liu L, Grulan JC (2007) Clay Assisted Dispersion of Carbon Nanotubes in Conductive Epoxy Nanocomposites. Advanced Functional Materials 17:2343-2348
[41] Kornmann XK, Lindberg H, Berglund LA (2001) Synthesis of epoxy–clay nanocomposites: influence of the nature of the clay on structure. Polymer 42:1303-1310
[42] Georgakilas V, Bourlinos A, Gournis D, Tsoufis T, Trapalis C, Mateo-Alonos A, Prato M (2008) Multipurpose Organically Modified Carbon Nanotubes: From Functionalization to Nanotube Composites. Journal of the American Chemical Society 130:8733-8740
[43] Ray SS, Okamoto M (2003) Polymer/layered silicate nanocomposites: a review from preparation to processing. Progress Polymer Science 28:1539-1641
[44] Zhang JP, Wang AQ (2009) Synergistic effects of Na+-montmorillonite and multi-walled carbon nanotubes on mechanical properties of chitosan film. eXPRESS Polymer Letters 3:302-308
[45] Geim AK (2009) Graphene: Status and Prospects. Science 324:1530-1534

[46] Lee JH, Shin DW, Makotchenko VG, Nazarov AS, Fedorov VE, Kim YH, Choi J-Y, Kim JM, Yoo J-B (2009) One-Step Exfoliation Synthesis of Easily Soluble Graphite and Transparent Conducting Graphene Sheets. Advanced Materials 21:1-5
[47] Geim AK, Novoselov KS (2007) The rise of graphene. Nature Materials 6:183-191
[48] Park S, Ruoff RS (2009) Chemical methods for the production of graphenes. Nature Nanotechnology 4:217-224
[49] Liang J, Wang Y, Huang Y, Ma Y, Liu Z, Cai J, Zhang C, Gao H, Chen Y (2009) Electromagnetic interference shielding of graphene/epoxy composites. Carbon 47:922-925
[50] Stankovich S, Dikin DA, Piner RD, Kohlhaas KA, Kleinhammes A, Jia Y, Wu Y, Nguyen ST, Ruoff RS (2007) Synthesis of graphene-based nanosheets via chemical reduction of exfoliated graphite oxide. Carbon 45:1558-1565
[51] Villar-Rodil S, Paredes JI, Martinez-Alonso A, Tascón JMD (2009) Preparation of graphene dispersions and grafene-polymer composites in organic media. Journal of Materials Chemistry 19:3591-3593
[52] Iijima S (1991) Helical Microtubules of Graphitic carbon. Nature 354:56-58
[53] Shokrieh MM and Rafiee R (2010) A Review of the Mechanical Properties of Isolated Carbon Nanotubes and Carbon Nanotube Composites. Mechanics of Composite Materials 46:155-172
[54] Suave J, Coelho LAF, Amico SC, Pezzin SH (2009) Effect of sonication on thermomechanical properties of epoxy nanocomposites with carboxylated-SWNT. Materials Science and Engineering A 509:57-62
[55] Coelho LAF, Pezzin SH, Loos MR, Amico SC (2009) General issues in carbon nanocomposites technology. In: Lechkov M and Prandzheva S (ed) Encyclopedia of Polymer Composites: Properties, Performance and Applications, 1 ed. Nova Science Publishers Inc, Hauppauge
[56] http://www.nanosperse.com/ Accessed May 2010
[57] Pizzutto CE, Suave J, Bertholdi J, Pezzin SH, Coelho LAF, Amico SC (2010) Mechanical and Dilatometric Properties of Carboxylated SWCNT/Epoxy Composites: Effects of the Dispersion in the Resin and in the Hardener. Journal of Reinforced Plastics and Composites 29:524-530
[58] Gojny FH, Wichmann MHG, Fiedler B, Kinloch IA, Bauhofer W, Windle AH, Schulte K (2006) Evaluation and identification of electrical and thermal conduction mechanisms in carbon nanotube/epoxy composites. Polymer 47:2036-2045
[59] Zhang W, Phang IY, Liu T (2006) Growth of Carbon Nanotubes on Clay: Unique Nanostructured Filler for High-Performance Polymer Nanocomposites. Advanced Materials 18:73-77
[60] Wang ZC, Xu CL, Zhao YQ, Zhao DD, Wang Z, Li HL, Lau KT (2008) Fabrication and mechanical properties of exfoliated clay-CNTs/epoxy nanocomposites. Materials Science and Engineering A 490:481-487
[61] Lan Y, Lin J (2009) Observation of Carbon Nanotube and Clay Micelle like Microstructures with Dual Dispersion Property. Journal of Physical Chemistry A 113: 8654-8659
[62] http://www.nanoclay.com/testdata.asp#mechanical Accessed May 2010
[63] http://www.sud-chemie.com Accessed May 2010
[64] http://www.elementis-specialties.com/ Accessed May 2010
[65] http://www.ptonline.com/articles/200411fa2.html Accessed May 2010
[66] http://www.graphistrength.com/sites/group/en/products/detailed_sheets/multi_wall_carbon_nanotubes_graphistrength/home.page Accessed May 2010

[67] http://www.nanocor.com/ Accessed May 2010
[68] http://www.hyperioncatalysis.com/ Accessed May 2010
[69] http://www.inmat.com/ Accessed May 2010
[70] http://www.cleantechnano.com/applications.html Accessed May 2010
[71] http://www.nanocompositesinc.com/ Accessed May 2010
[72] http://www.csiro.au/files/files/p55a.pdf Accessed May 2010
[73] Zhang M, Atkinson KR, Baughman RH (2004) Multifunctional Carbon Nanotube Yarns by Downsizing an Ancient Technology. Science 306:1358-1361
[74] http://www.vorbeck.com/ Accessed May 2010

Abbreviations

VGCF - vapor-grown carbon fiber
CNTs - carbon nanotubes
CNSC - Carbon NanoSphere Chain™
NGP - nanoscale graphene platelets
SWCNTs - single-walled carbon nanotubes
DWCNTs - double-walled carbon nanotubes
MWCNTs - multi-walled carbon nanotubes
PA 6 - polyamide 6
MMT - montimorrilonite
PMMA – polymethylmethacrylate
MMT-Na^+ - Sodium modified montmorillonite clays
PS – polystyrene
PC – polycarbonate
PVA - polyvinyl alcohol
SMC - sheet molding compound
CVD - chemical vapour deposition
FGS - functionalized graphene sheets
GO - graphite oxide
GS - graphene sheets
ESD - electrostatic discharge

9 Nanomaterials for Applications in Refractory Materiais

Álvaro Niedersberg Correia Lima and Diogo Kramer Topolski

Departamento de Engenharia de Materiais, Universidade Federal do Rio Grande do Sul, 90035190, Porto Alegre, Brazil
E-mail: topolski@gmail.com

Abstract. Refractories based on nanostructured materials are emergent technological products in several applications in the steel and cement industries. To better understand the nanostructured materials in refractories applications this chapter shows a general perspective of the use of these materials in industrial refractories. Refractories research and development based on conventional technical concepts and techniques seem to have been conducted exhaustively. If refinement of refractory technology is to be expected, the introduction of new technologies is indispensable. Nanostructured materials improve several properties of refractory materials like mechanical resistance, thermal conductive, carbon oxidation protection.

Keywords: refractory, cement castables, nanostructured.

9.1 Introduction

Refractory materials are inorganic chemical substances, single or polyphase in nature, which are processed at high temperature and/or are intended for high-temperature applications. They are employed wherever a process involves treatment or exposure to elevated temperatures, e.g. the melting of metals, the sintering of ceramics, the melting of glass, the processing of hydrocarbons and other chemicals, etc [1]. These ceramics are the raw materials of the industrial refractories.

With the exception of certain types of fusion-cast refractories, almost all refractories involve an aggregate of small particles. They are made from a formed material, which has a very wide particle-size distribution, from several micrometers (fine particles) to several millimeters (coarse particles) across. Unlike fine ceramics, which are made by subjecting a formed material consisting of particles of almost uniform size to an extremely large firing shrinkage for densification of 30 – to 50 wt% (because of this large change in volume, it is difficult to manufacture large ceramic products in bulk), refractories, which effectively utilize the wide particle-size distribution, are free from such large firing shrinkage even when the formed material is subjected to high-temperature firing/hysteresis for the densest packing or pore-shape control. In addition, the structure of a refractory has a greater effect on the properties of the refractory than composition control [2, 3].

Ordinarily, in the field of conventional refractories, it is fine particles several micrometers in size that are routinely subjected to blending control. This paper describes the development of a "nano structural matrix"— an original technology whereby a nano region is established in an ordinary matrix. For the purpose of this development, nanotechnology was positively applied [4].

9.2 Industrial Application

9.2.1 Nanostructured Oxides

9.2.1.1 Sintering Aid

The nanostructured materials have a very high sintering activity. This is caused because a great part of the atoms are on the particle surface [5]. These characteristics turn nanostructured materials into a good sintering aid, decreasing the refractories firing temperature.

Tontrup [6] used Aeroxide ® TiO_2 P25 (properties shown in Table 9.1) as sintering aid in 99% alumina refractory bricks. Using just 0.4% of it, the firing temperature was reduced by more than 200ºC (1730ºC to 1500ºC), increasing almost 30% of the cold crushing strength (CCS) – 8.5 MPa to 13.8MPa – compared to a conventional brick.

Table 9.1. Properties of Aeroxide ® TiO_2 P25.

Properties	Aeroxide ® TiO_2 P25
Specific surface area (BET)	90 m^2/g
Primary particle size	14 nm
Crystal structure	> 95% Anatase

9.2.1.2 Reinforcing Additive

Tontrup [6] also used Aeroxide ® AluC (properties shown in Table 9.2) in industrial corundum sliding gates formulation. The CCS increased 27% with an addition of 0.25%.

Table 9.2. Properties of Aeroxide ® AluC.

Properties	Aeroxide ® Alu C
Specific surface area (BET)	100 m^2/g
Primary particle size	13 nm
Crystal structure	2/3 γ and 1/3 δ

The same material also improved the hot crushing strength (HCS) of the shaped corundum material. The HCS increased 17% at 1400ºC while the addition of clay decreases HCS by 15%.

Aeroxide ® AluC showed good qualities in carbon bonded alumina bricks too. The reinforcement on the carbon matrix cooked at 1000ºC is so great that the CCS increased 64%with 0.4%.

Aeroxide ® AluC also increases the matrix adhesion of industrial magnesia-spinel refractories, one of the most promising products in the refractory industry. Table 9.3 shows the thermal shock resistance of some specimens of these refractories.

Table 9.3. Thermo shock resistance of magnesia-spinel refractories.

Additive	CCS (MPa)	Thermo shock resistance (cycles)	Cracks after number of cycles
0.00%	63.1	14	4
0.1% Aeroxide ® AluC	66.0	18	8
0.4% Aeroxide ® AluC	67	14	4

With the addition of 0.1% of Aeroxide ® AluC, the thermal resistance showed great improvement. This effect is also found in magnesia-chrome refractories with combination of nanostructured TiO_2 and Al_2O_3.

Aeroxide ® AluC not only improves the properties of solid bricks. Trontup [6] also tested the strength of dried alumina castables. With an addition of 0.3% of this material, the CCS increased 41%. This has a potential to reduce cement content, because the nanostructured oxides provide strong adhesion in dried refractories because of their high surface area. In fired refractories, the major effect is reinforcing the binding matrix.

9.2.2 Utilization of Nanostructured Materials in Castable Cement

Calcium aluminate cement (CAC) is the most used hydraulic binder in refractory castables. However, the presence of CaO in the composition can be deleterious to the refractoriness of some ceramic systems, such as those containing microsilica. Moreover, CAC-based castables impose some special requirements concerning their curing and drying condition that could extend the processing time [7]. In this context, colloidal silica has been pointed out as an alternative for a calcium-free binder agent for refractory castables [8] and [9].

Colloidal silica is a stable dispersion of nanosized particles of amorphous silica. The different production routes available can lead to sols with distinct solids concentration, particle size and shape, and pH [10]. Colloidal silica particles can be linked together using different setting mechanisms, such as gelling and coagulation, providing initial strength when applied to ceramic systems. Concerned to gelling, the interaction of silica particles triggers the siloxane bonding (Si–O–Si) and the build-up of a three-dimensional network. With regard to coagulation, an additive (usually electrolytes) bridges the particles causing close-packed clumps. Both setting mechanisms are influenced by pH change, particle size and its concentration, the presence of electrolytes and organic liquids, and temperature [10]. Besides its binder effect, colloidal silica is also reported as a dispersant for ceramic powders [10] and [11].

Anderson [12] compared two similar cement (70% alumina) and colloidal silica bonded castables to Magnec/Metrel Inc. The porosity of the colloidal silica bonded remained constant (14.5%) with the increase of firing temperature (from 50ºC to 1400ºC). In water bonded, the porosity increased more than 4% in the same temperature range. The lower porosity at intermediate temperatures leaves the colloidal silica bonded refractory more resistant to oxidation, gas attack and slag attack.

Cement bonded materials often have greater initial strengths than colloidal silica bonded materials due to the hydraulic bond. However, CAC materials lose this strength as dehydration takes place. A cold modulus of rupture test showed a dip in the strength curve at about 500ºC for a cement bonded material due to a hydration phase loss and the associated bonding loss. Although the colloidal silica bonded product showed a lower initial strength, it showed a good progression in strength development and quickly equaled the cement bonded material at moderate temperatures (< 500ºC).

At elevated temperatures, the strength differences become more dramatic. Low-melting-temperature CaO-Al_2O_3-SiO_2 phases associated with CAC castables are responsible for liquid phase formation at the indicated temperatures and results in lower hot strengths. Because the colloidal silica bonded materials are CaO-free, they do not generate these low melting phases and typically exhibit higher hot strengths which results in better in-service erosion resistance for these materials.

Glassy phase formation and volume instability associated with CAC castables often cause degradation of properties upon thermal cycling. In most systems, colloidal silica bonded products have been shown to be superior in cycling applications. Anderson shows the strength loss of a CAC castable and a similar colloidal silica bonded pumpable material. After only five cycles, the CAC material shows a loss of over 85% of its initial strength, while the colloidal silica bonded material loses strength much more slowly and at a steady, predictable rate (after 15 cycles loss 50%) [12].

Colloidal silica bonded refractories have exhibited excellent flow behavior, hot strength, thermal shock resistance, oxidation and acid resistance, and bond strength in a variety of applications [13].

9.2.3 Carbon Black Utilization in MgO-C Refractories

In terms of the added nanocarbon particles, there are two types of carbon black — the single-sphere and aggregate types (with large grain-size distribution). The single sphere provides bigger densification and improves the erosion resistance. The aggregate type decreases the elasticity, promotes stress relaxation, pore segmentalization and improves thermal shock resistance and insulation [14].

Tamura *et al.*[14] manufactured a ultra extra-low carbon MgO-C refractory with a nanostructural matrix aiming for 3% total carbon, including the carbon contained in the organic binder used. Refractory specimens with the mixing ratios of single-sphere and aggregate varied as shown in Table 9.4 and added to a suitable amount of binder were prepared. The principal characteristics of the types of carbon black utilized are shown in Table 9.5.

Table 9.4. Experimental condition of experimental brick (%)

Specimen	CB1	CB2	CB3	CB4	CB5
MgO	95	95	95	95	95
Single spheres	2	1.5	1	0.5	0
Aggregate	0	0.5	1	1.5	2

Table 9.5. Characteristics of carbon black

	Single sphere	Aggregate
Specific surface area (BET) (m^2/g)	19	60
Oil absorption (ml/100g)	30	130
Component particle diameter (nm)	90	40

The addition of single-sphere alone (2%) shows a marked tendency to increase both strength and elasticity. When the mixing ratio of the aggregate is 0.5% or more, the modulus of elasticity decreases to about half, although the strength declines slightly. In this case, therefore, it can be expected that the thermal shock resistance (spalling resistance) will improve.

On the basis of the above results, Tamura *et al* made some experimental bricks (X and Y in Table 9.6).

Table 9.6. Effect addition of agregate types (%).

Specimen (MgO-C(3%))	X	Y	Z (conventional)
MgO	96	96	78
Single sphere	1.0	0	0
Aggregate	0.5	0	0
Aggregate B	0	1.5	18
Flake graphite	0	0	18
Resine binder	2.5	3.2	2.8
Spalling test (*1) cycle	12	15	15

(1*) cycles until peeling occurs by repetition of dipping specimen in molten iron at 1400ºC

Specimen X was prepared by adding single-sphere type carbon black, which effectively increases the refractory's density, and aggregate carbon black—which improves the thermal shock resistance of the refractory. Carbon black b used in Specimen Y is a hybrid of single sphere and aggregate. Specimen Z, a conventional brick available on the market, was used for comparison. The total carbon in X and Y, respectively, is 3% (flaky graphite 0%, carbon black 1.5%), whereas that of Z is 20% (flaky graphite 18%), more than six times greater.

In a spalling test, specimens X and Y showed spalling resistance comparable to that of Specimen Z, despite the fact that the carbon content of X and Y was only about one-sixth that of Z. Thus, using carbon black appropriately, it is possible to impart excellent spalling resistance to the brick. The heat conductivity of Specimens X and Y, which contain no graphite, is about one-seventh that of Specimen Z and hence, X and Y display high thermal insulation performance [15].

The results of a corrosion resistance test of the specimens carried out on the slag line of an actual ladle. The wear rates of Specimens X and Y containing 3% carbon are low when compared with that of Specimen Z.

All these results show the improvement caused by nanostructural carbon black in the matrix. Besides the improvement on the bricks properties, the lower use of carbon generates other benefits. Using a refractory which contains a considerable proportion of carbon in a converter is problematic because of the large amounts of oxygen blown into the converter. Also, with an eye on the necessity of energy saving, conserving natural resources and tackling environmental problems in the future, it is considered absolutely imperative to reduce the amount of graphite added and to develop low-carbon refractories [16].

References

[1] Moore RE (2001) Refractories, Structure and proprieties of, In: Buschow KHJ (ed.) Encyclopedia of Materials: Science and Technology, Pergamon, Oxford

[2] Fujihara S, Tamura T, Ikeda M, Nakai M (1989) Development of Carbon Blacks. Nippon Steel Technical Report 41:1-6

[3] Sugita K, Ikeda M, Tamura S (1984) Microstructural improvements in slide gate plates for steel pouring. Am Ceram Soc Bull 63:886-889

[4] Ochiai T (1999) Application of Microwave Drying to Monolithic Ladle Linings. Journal of the Technical Association of Refractories 19:31-45.

[5] Edelstein AS, Cammarata RC (1996) Nanomaterials: synthesis, properties and applications. Physics Publishing, London

[6] http://www.aerosil.com/product/aerosil/de/industrien/anwendungsbereiche/leuchtmittel/Page45docs/default.aspx. Accessed August 2010

[7] Cardoso FA et al (2004) Effect of curing conditions on the properties of ultra-low cement refractory castables. Refract Appl News 9:12–16

[8] Banerjee S (1988) Monolithic Refractories: a Comprehensive Handbook, World Scientific/The American Ceramic Society

[9] Banerjee S et al (1992) Electrochemical Heat Source. US Patent 798,347

[10] Iler RK (1979) The Chemistry of Silica: Solubility, Polymerisation, Colloid and Surface Properties and Biochemistry. John Wiley & Sons, Hoboken.

[11] Zhu X et al (2001) Dispersion properties of alumina powders in silica sol. J Eur Ceram Soc 21:2885–2897
[12] http://www.ceramicindustry.com/Articles/Feature_Article/0c2b3fcb0cac7010VgnVCM100000f932a8c0 Accessed August 2010
[13] Anjos RD et al (2008) Workability and setting parameters evaluation of colloidal silica bonded refractory suspensions. Ceramics International 34:165-171.
[14] Tamura S et al (2008) Technological Philosophy and Perspective of Nanotech Refractories. Nippon steel technical report 98:18-28.
[15] Takanaga S, Ochiai T, Tamura S, Kanai T, Nakamura H.: (2003) Development of the Nano Structural Matrix. Proceedings of UNITECR2003, p.521-524
[16] Ikeda, M et al (2005) Nippon Steel Chemical Carbon Co. Bulletin of Chemical Daily 35:9-14

Abbreviations

BET - Specific BET (Braunauer, Emmet and Teller) surface area
CAC - Calcium aluminate cement
CCS - Cold crushing strain
HCS - Hot crushing strain

10 Materials for Adsorbent Applications

Fernando Machado Machado and Carlos Pérez Bergmann

Departamento de Engenharia de Materiais, Universidade Federal do Rio Grande do Sul, 90035190, Porto Alegre, Brazil
E-mail: fernando.machado@hotmail.com.br

Abstract. Nanoscience and Nanotechnology suggest that many of the current problems involving energy storage and industrial effluents could be solved or greatly ameliorated using the nano-scale adsorbents, called nanoadsorbents. Innovations in the development of novel materials, such as dendrimers, zeolites, nanocomposites based on metal hydrides and carbonaceous structures like carbon nanotubes are among the most exciting and promising. This chapter provides an overview of the use of nanomaterials as adsorbents. We highlight recent advances in the development of new nano-scale materials and processes for H_2 adsorption and adsorption of toxic metal ions, synthetic dyes, biological contaminants and acids in aqueous solutions.

Keywords: adsorption, nanomaterials, nanoadsorbents, carbon nanotubes, dendrimers, zeolite.

10.1 Introduction

Nanoscience and Nanotechnology have aroused considerable interest, not only because of the expectation of the impact nanomaterials may have on improving the quality of life, but also for the preservation of the environment. In recent years, nanotechnology has provided different types of nanomaterials that are being evaluated as possible nanoadsorbent materials, for example, dendrimers, zeolites, nanocomposites based on metal hydrides and carbonaceous structures like carbon nanotubes (CNT) [1,2] (see Figure 10.1). These have a range of physical and chemical properties that make them fascinating for the development of devices for energy storage (for example, adsorption of H_2) [3,4,5,6,7,8,9] and the development of absorbent devices with high sensitivity, selectivity and high efficiency for removal of heavy metals [10,11,12,13,14,15,16,17], synthetic dyes [18,19,20,21], biological contaminants [22,23] and acids [24] in aqueous solutions.

This chapter provides an overview of the H_2 storage and removal of contaminants from water using nanomaterials as adsorbents. The chapter discusses briefly the peculiar properties of each of these materials that contribute to their being used in such applications, followed by examples of studies on nanoadsorbents available in the literature at the moment.

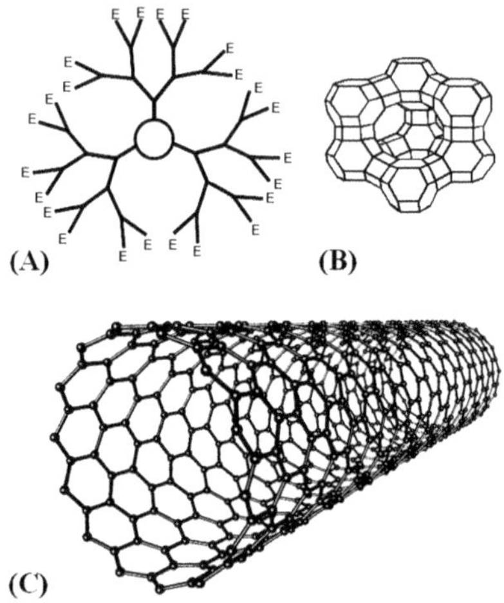

Fig. 10.1. Typical structures of nanomaterials with potential applications for adsorption: (A) dendrimer, (B) zeolite and (C) carbon nanotube.

10.2 Adsorption

The adsorption phenomenon is the mass transfer of a solute (adsorbate) present in a fluid phase to the porous surface of a solid phase (adsorbent) through the intermolecular attraction between them. If there is no kind of attraction, the molecules of adsorbate incident will collide elastically with the solid one and will return to the medium with the initial speed [25,26]. Adsorption is a process of interfaces in which a film of adsorbates is created on the surface of the adsorbent material. Analogous to surface tension, adsorption is a consequence of surface energy. Inside the adsorbent material, all connections required (whether covalent, ionic or metallic) of the constituent atoms of the material are filled by other neighboring atoms. However, the atoms of the adsorbent surface are not completely surrounded by other adsorbent atoms, i.e., they are associated with a smaller number of neighboring atoms. As the binding energy decreases the potential energy of the system, the surface has extra energy, called surface energy. Thus, the surface atoms are more reactive and therefore can attract adsorbates. This effect is more pronounced in nanostructured materials due to the large fraction of atoms on the surface.

The exact nature of the connection of the adsorbent–adsorbate depends on the species involved, but the adsorption process is generally classified as physical adsorption (physisorption) and chemical adsorption (chemisorption) [27]. In chemisorption, the forces of chemical attraction, typically covalent bonds, act between the surface of the adsorbent and adsorbate. Thus, there is a chemical combination between the surface of the adsorbent material and adsorbate, where electrons are shared and/or transferred and new electronic configurations can be formed through this process. On the other hand, in physisorption, physical adsorption forces, weak van der Waals or electrostatic forces, are the ones that work without there occurring any sharing or transferring electrons and with the adsorbate maintaining its identity, although it may be deformed by the presence of surface force fields.

10.3 Morphology of Nanostructured Particles

The nanostructured materials exhibit a very large number of atoms, or chemical species in general, likely to reduce the free energy of the system that is inserted through chemical bonds or physical interactions with other chemical species present in their vicinity. The morphology of these particles can vary significantly by modifying the ability to serve as adsorbent for chemical species in the reduction of free energy of the system. Examples of materials with such typical morphologies are dendrimers, zeolites and carbon nanotubes.

10.3.1 Dendrimers

Dendrimers are formed by polymer molecules, i.e., by units (monomers) that are repeated, which form branches like trees in a process of self-aggregation. The dendritic architectural class can be divided into four subclasses: random hyperbranched polymers, dendrigraft polymers, dendrons and dendrimers [28]. Currently, more than 100 different dendrimers and 1,000 chemical surfaces are known. The dendritic structure resulting is highly branched, symmetrical and has a large number of functional groups on the surface. This structure consists of three basic components (Figure 10.1(A)) [29]: a multifunctional core (○) where other molecules can be trapped [30], branching units (Y) that emanate from the central nucleus and external-capping groups (E) [31]. Their unique structure and small size (between 10 and 20 nm) make them particularly attractive for use as high capacity nanoadsorbents for water purification. Dendrimers can trap, for example, toxic metal ions in their structure, which can be subsequently removed by processes of ultrafiltration (UF) [32].

10.3.2 Zeolites

A zeolite is a crystalline aluminosilicate with a structural frame including large cavities occupied by cations and water molecules, both having considerable freedom of movement, permitting ion exchange and reversible dehydration (see Figure 10.1(B)) [33]. The removal of water molecules and the replacement of exchangeable cations do not alter the basic structure of zeolites.

Although conventional approaches report the synthesis of zeolite production on a micro-scale, highly uniform crystals of zeolites on the nano-scale can be synthesized with dimensions ranging from 10 to 500 nm [34,35,36]. Zeolite on the nano-scale has great mechanical and chemical resistance [37] and high surface area (400-1000 m^2/g) [1,38,39] and high density of adsorption sites. Studies have reported that zeolites on the nano-scale present higher adsorption capacity than those on the traditional micro-scale [38], which make them attractive to immobilize in its structure, for example, heavy metals from acid mine effluent [17] and pollutant molecules, such as toluene and nitrogen dioxide [38,39].

10.3.3 Carbon Nanotubes

CNT is an allotrope composite of carbon, consisting of sp^2 hybridized carbon atoms with bonds, distorted due to the curvature of the structure, which can be found in the form of a single wall (single-walled carbon nanotube - SWNT) (Figure 10.1(C)) [40,41] or multiple walls (multi-walled carbon nanotube - MWNT) [42]. These can be considered as one-dimensional nanostructures (1-D) because of their small diameter and their large length/diameter ratio. The CNTs produced are unfortunately insoluble in many liquids, for example, in aqueous solutions, and polymer resins in most solvents [43,44]. To facilitate and uniform the dispersion of CNTs in liquid it is necessary to attach functional groups or molecules to their walls, without significantly altering their properties as desired.

With regard to the adsorptive properties of CNTs, one must consider their internal and external surfaces. Moreover, the curvature of the graphene sheet can result in a lower heat of adsorption compared to that of a flat sheet. In fact, the rolling up of the graphene sheet to form the nanotube causes a rehybridization of carbon orbital (non-planar sp^2 configuration), thus leading to changes of the π-density of the graphene sheet [45].

Diverse studies highlight the porous nature of the CNT [45,46,47,48,49]. Pores in aggregated MWCNTs can be mainly divided into inner hollow cavities of small diameter (ranging from 3 to 6 nm in a narrow distribution of sizes) and aggregated pores (widely distributed, 20 to 40 nm in), formed by the interaction between isolated MWCNTs [49]. The aggregated pores are formed by entanglement of tens and hundreds of isolated MWCNTs, which are adhered to each other as a result of van der Waals forces.

Many methods of characterization have shown the microporous nature of SWCNTs and the mesoporous nature of MWCNTs [45]. Thus, the former often exhibit a specific surface area higher than the others. Experimentally, it is reported that the surface area of as-grown SWCNTs ranges between 400 and 900 m^2/g (micropore volume of 0.15-0.3 cm^3/g) and as-grown MWCNT has a surface area between 200 and 400 m^2/g (mesopore volume of 0.5-2) [45]. For SWCNT, the diameter of the tubes and the number of tubes in the bundle will affect mainly the surface area value [45].

As seen in Figure 10.2, adsorption in SWCNT can occur at four different sites, these being: the grooves formed at the contact between adjacent tubes, on the outside of the bundles, the interstitial channels between the tubes in bundles, the inside of the nanotubes (pores) with open ends [45, 50, 51].

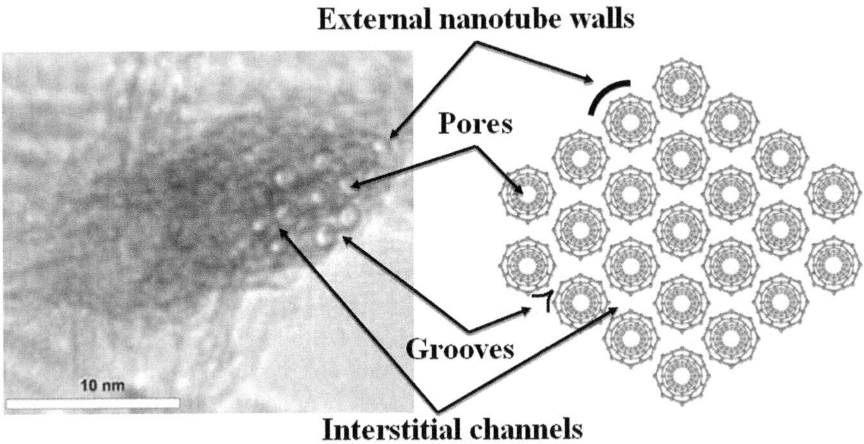

Fig. 10.2. TEM image of the SWCNT bundle (left) and schematic model for the possible adsorption sites of the SWCNT bundle (right).

The interstitial channels formed between the nanotubes have good potential for the adsorption of small adsorbates, for example, gases [51].

In MWCNTs the adsorption sites consist of pores aggregated (Figure 10.3), in the inner region in MWCNT with open ends and on the external walls [49]. In this last case, the presence of defects, such as an incomplete graphene sheet, should be considered. The pores aggregated play a more significant role in the adsorption of large biological contaminants such as bacteria and viruses [22].

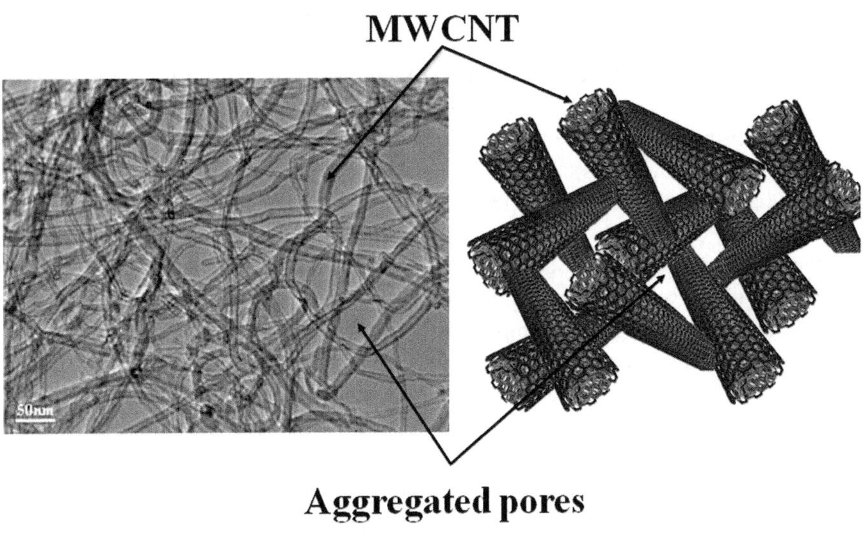

Fig. 10.3. TEM image of the aggregated MWCNTs (left) and schematic model for the pores aggregated of MWCNT (right).

10.4 Hydrogen Storage

Hydrogen adsorption on carbon-based materials is the amount of hydrogen that remains near the surface of the solid carbon due only to physical forces — van der Waals interactions — of the carbon atoms on the hydrogen molecules (process of physisorption) [8,52].

Carbon is known as one of the best adsorbents for gases [53]. This property is based on: (i) morphology of carbonaceous structures in the form of a fine powder with high porosity and (ii) the existence of specific interactions between the carbon atoms and the gas molecules. As examples of carbon-based materials, as the leading candidates for H_2-absorbent materials, the following could be cited: activated carbon, fibers (nano) and carbon nanotubes. A survey of State of the Art H_2 adsorption shows that the values of adsorption obtained by different authors are divergent [54,55]. The storage of H_2 in carbon structures has generated controversy in terms of efficiency, reproducibility of experiments and theories to explain the phenomena involved. However, the National Renewable Energy Laboratory - NREL, USA, works with the expectation of up to 10% by weight for technologies based on hydrogen adsorption by carbon-based materials. H_2 adsorption on activated carbons was one of the first possibilities investigated. This is due to the high specific area of these materials, allowing the gas adsorption in microporous (mesoporous and macroporous do not influence the amount of adsorbed gas). However, the adsorption of H_2 in this type of carbonaceous structure was not feasible for practical application: at room temperature, the gravimetric energy density of the system is very small. The most interesting results for commercial activated carbons ranged around 0.5% by mass, at a pressure of 60 bar, although the literature shows values of up to 8% by mass with 50 bar at 165 K. Graphite also has the capacity to adsorb H_2 on its surface. When ground below 200 microns, kept in the presence of H_2 for 80 hours, it was able to adsorb 7.4 wt.% at a pressure of 10 bar and a temperature of 300 K. The explanation for this surprising performance was the generation of metastable structures, including new forms of carbon structures, like — onions (nanonions).

Exfoliated graphite arises from chemical treatment and subsequent heating of the graphite. The exfoliated structure of carbon blades generates an increase in volume of the material. The highest adsorption capacity of H_2 by exfoliated graphite when compared to conventional graphite, would reside in that lamellar separation. Figure 10.4 shows SEM images of samples of conventional and exfoliated graphite. The term carbon nanofibers (CNF) summarizes a large family of carbon nanofilaments. These can be differentiated according to the arrangement of the layers of graphite or the morphologies and growth modes. Regarding the possibility of H_2 adsorption, the spatial structure of CNFs has advantages due to the accessibility of the entire outer blade and diffusion for short trips inside the nano structure.

CNTs have also been investigated for use in H_2 storing [3,4,5,6,7,8]. The adsorption of hydrogen can occur in sites located both on the inside as on the outside walls of the CNTs [8]. Through Raman spectroscopy it is possible to detect

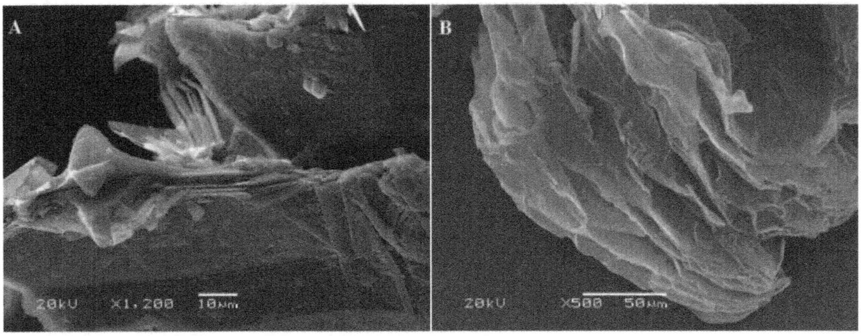

Fig. 10.4. SEM image of conventional (a) and exfoliated (b) graphite.

hydrogen in the empty space of CNTs in the form of H_2, which was confirmed by molecular dynamics simulations. In the case of long SWCNTs closed with fullerenes in the end, hydrogen can only access the inner tube through the formation of hexagonal carbon atoms in the tube wall. An open tube without the presence of fullerene on the tip, allows easier access of the H_2 molecule inside the tube [8]. The adsorption of H atoms in MWCNTs is favored in the outer walls of concentric tubes, where the storage capacity is independent of diameter. For this conception of adsorption, the SWCNTs are better means of storage of H_2 than MWCNTs [8]. The hydrogen adsorption capacity of CNTs can be increased by milling, which promotes the opening tip and surface defects [56].

Regarding the importance of the adsorption capacity of CNTs for a reliable storage, their volume density must be taken into consideration. The amount of H_2 stored is directly proportional to the adsorption expressed in weight% and, consequently, the volume density of the adsorbent. For CNT, the volume density is about 20 kg/m^3, while for other conventional, carbon-based adsorbents it is of the order of 500-1000 kg/m^3. Mathematical simulations of hydrogen–carbon interaction [52,57] have given rise to the expectation that the adsorption of H_2 can be optimized by changing the diameters of CNTs and the space between the bundles, as well as by its functionalization or doping, using metals like titanium [7,58]. These results are important indicators for further development of the CNT-making process. Figure 10.5 shows curves of H_2 adsorption of different materials based on carbon, investigated comparatively.

Although they have different structures, the adsorption curves of carbon-based materials are very similar. Apart from the higher adsorption capacity of CNTs, the results of H_2 adsorption on activated carbon and exfoliated graphite, with significant values of H_2 adsorption, are interesting because of the low cost of those absorbers. In the comparative H_2 adsorption, one must consider sources of error such as quantity, purity and homogeneity of the adsorbent and techniques for the determination of adsorption. These factors lead to low reproducibility of results. The experimental errors can be associated with leakage of H_2 or pressure change due to varying temperatures and the return to thermal equilibrium after pressurization.

Another possible form of hydrogen storage is the use of metal hydrides [9], since these compounds can store a relatively large amount of H_2 in solid state at room temperature and pressure [59], providing easy and safe handling. Mg-based

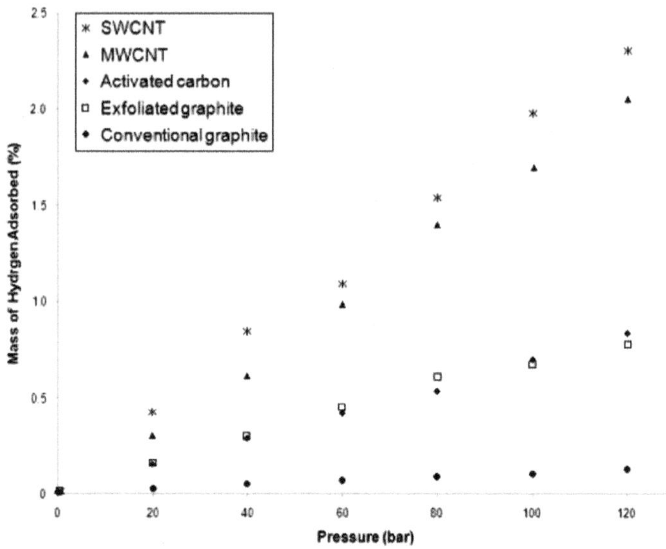

Fig. 10.5. H_2 adsorption of different materials based on carbon.

alloys are relatively lightweight, inexpensive, and are promising materials for hydrogen storage due mainly to high gravimetric and volumetric capacities of their hydrides [60]. Among these alloys, Mg_2FeH_6 has the highest storage density by volume (150 g H_2/l, which is 9.1×10^{22} atoms H/cm^3), i.e., twice the storage density of liquid, and its hydrogenation reaction is reversible and has a high cyclic stability [61] while MgH_2 has the highest gravimetric storage capacity (7.6 wt.% H). In the latter case, if MgH_2 is nanostructured, the kinetics of absorption/desorption are improved [62].

As seen above, hydrogen storage is a key issue for success and achievement of hydrogen economy and technology. At the moment, it appears that the stage of development of most of these technologies is very embryonic, but it can eventually lead to advances in storage of H_2 in the coming years, mainly for the use in fuel cells for automotive, mobile phones and laptop applications.

10.5 Adsorbents for Wastewater

It is estimated that today about 1.1 billion people lack access to potable water and approximately 35% of the inhabitants of the "Third World" die of diseases linked to water quality. If today only a few countries have a shortage of drinking water supply, it is estimated that by 2025, based on extrapolation of current data, more than half the countries of the world will be in such a crisis [2]. Although unpleasant, scientific studies strongly indicate that many of today's water quality problems can be solved with the use of nanotechnology [2].

Due to its characteristic structure on the nano-scale (see Figure 10.1 (A)), dendrimer as nanoadsorbent has been tested in water purification [63].This material can trap toxic metal ions present in aqueous solutions on its nanostructure [32,64] which are then removed by UF process, called dendrimer-enhanced ultrafiltration

- DEUF [32]. This technique was first tested successfully using dendrimers poly (amide-amine) PAMAM dendrimers with an ethylenediamine (EDA) core and terminal NH_2 groups to remove Cu(II) from aqueous solutions [32].

On a mass basis, the binding capacity of Cu(II) with PAMAM dendrimers is much larger and more sensitive to solution pH than those of linear polymers with amine groups. The separation of the whole dendrimer/Cu(II) solution can be achieved by UF with the appropriate molecular weight cut-off MWCO. Dendrimers loaded with metal ions can be regenerated simply by decreasing the solution pH to 4, thus allowing the recovery of Cu(II) and recycling of dendrimers [32].

Zeolites have been shown to be effective adsorbent for a range of adsorbates [1]. Zeolite, such as the type NAP1 ($Na_6Al_6Si_{10}O_{32}$, $_{12}H_2O$) has a high density of sites of ion exchange of Na and they can be synthesized by hydrothermal activation of fly ash with low Si/Al ratio at 150°C in 1-2 M NaOH solutions [17]. Zeolite NAP1 was evaluated as a medium of ionic exchange for removing heavy metals from acid mine effluent [17]. Studies also reported the successful use of synthetic zeolites NAP1 with adsorbents of Cr(III), Cd(II), Ni(II), Zn(II) and Cu(II) wastewater of galvanized metal [65]. Self-assembled monolayers on mesoporous supports (SAMMS) are providing opportunities for developing more effective adsorbents for toxic metal ions [66] and anions [67]. These materials are produced by surfactant-templated synthesis of mesoporous ceramics. This produces nanoporous ceramic oxides with a surface area of approximately 1000 m^2/g and high density of adsorption sites, which can be functionalized to increase their selectivity.

Among the nanomaterials with potential applications in adsorption processes of contaminants in aqueous effluents, CNTs are those that have gained greater prominence. They have attracted growing interest as a new nanoadsorbent for water treatment and wastewater containing heavy metal ions such as Zn(II) [10], Ni(II) [11], Cd(II) [12], Pb(II) [13], Cr(VI) [14,15] and Cu (II)[16], synthetic dyes such as Direct Yellow 86 and Direct Red 224 [18], methylene blue [19] and Procion Red MX -5B [20], biological contaminants [22,23] and acids [24]. They are an attractive alternative for removal of organic and inorganic contaminants from water, due to their high aspect ratio, large surface area that can still be modified. Recently, CNTs have proven to be efficient adsorbents with a capacity that exceeds that of activated carbon [10,15,68].

For divalent metal ions (Cd(II), Cu(II), Ni(II), Pb(II), Zn(II)), parameters such as total surface oxidized, the pH and temperature play a fundamental role in CNT sorption [69]. Research shows that the adsorption capacity increases remarkably in CNT after oxidation [10,12,16]. The ability of adsorption is also dependent on the pH of the solution. For the Cu(II) adsorption, for example, adsorption on oxidized MWCNTs increases significantly at pH = 6 [16]. As for the ions Zn(II) and Ni(II) adsorption capacity increases with increasing pH in the range between 1 and 8, reaching a maximum adsorption at pH between 8 and 11, decreasing from pH=12 [10,11].

The temperature at which the adsorption process is carried out will influence both the rate of sorption as the degree in which sorption occurs. The adsorption capacity, for example, of Zn(II) and Pb(II) increases considerably with an increase in temperature [10,13], indicating an endothermic reaction. The time required for the adsorption of about 50% of maximum sorption capacity of Zn(II) is reached faster at higher temperatures [10]. The mechanisms by which metal ions are adsorbed on CNT prove to be quite complex and can be attributed to: electrostatic attraction, precipitation, adsorption and chemical interaction between metal ions and functional groups anchored to the surface of CNTs [69]. It is believed that the chemical interaction between metal ions and functional groups of the surface of CNT is the major mechanism of adsorption [10,11].

The recovery of metal ions and regeneration of carbon nanotubes can be easily achieved by changing, with the aid of acids or bases, the pH of the solution, without significantly harming the performance of the CNTs, indicating that CNTs are promising adsorbents for the treatment of effluent [10,13].

The removal of hexavalent chromium (Cr(VI)) from aqueous solutions using functionalized (oxidized) and non-functionalized MWCNTs has also been studied [14,15]. The removal of Cr(VI) on oxidized MWCNTs is favored in solutions with low pH (with maximum removal at pH <2 [14]). The main removal mechanisms of Cr(IV) are identified as: intraparticle diffusion, electrostatic interactions and ion exchange mechanism (redox reactions) [14]. The latter appears as the main mechanism for adsorption of chromium by MWCNTs. The redox reaction between the adsorption of Cr(VI) and the surface of oxidized MWCNTs leads to the formation of Cr(III), and subsequently, the sorption of Cr(III) on MWCNTs (MWCNT exists on the surface of both Cr(VI) and Cr(III)). Removal of Cr(VI) is effective and without significant release of Cr(III) to the environment. The non-functionalized MWCNTs show a higher adsorption capacity than the functionalized MWCNTs [15]. This is due to the fact that the functionalized MWCNTs contain electron-rich atoms in their functional groups which repel the negatively charged dichromate ions that inhibit the adsorption process. Both functionalized and non-functionalized MWCNT possess an adsorption capacity higher than that of activated carbon [15].

CNTs have proven to be an alternative for the adsorption of direct dyes, for example, and Direct Yellow 86, Direct Red 224 [18] and cationic dyes such as methylene blue [19]. Based on the average free energy of adsorption, the suggestion is that adsorption of direct dyes in CNTs can be characterized as an ion exchange process [18]. The analysis indicated that the thermodynamics of adsorption of both direct dyes and the cationic dyes were endothermic and spontaneous [18,19]. In addition, the adsorption of both dyes occurs through the physisorption process [18,19]. The adsorption of reactive dyes like Procion RED MX-5B (CI Reactive Red 2) in MWCNTs has also been considered [20]. The results suggest that the adsorption of reactive dye in MWCNT decreases with increasing pH, but increases with increasing temperature. As with direct and reactive dyes, the adsorption of Procion RED MX-5B is given through the physisorption process and spontaneously.

CNTs show high adsorption capacity for removal of a wide variety of biological contaminants like bacteria, viruses, natural organic matter (NOM) and cyanobacteria from aquatic systems [22,23]. The high capacity of adsorption of CNTs in comparison with other adsorbents is mainly attributed to its fibrous form with high aspect ratios and large surface area that can be easily accessed by biological contaminants, and to the presence of well-developed mesopores [23].

Adsorption of fulvic acid by CNT has been the subject of studies since [24]. Adsorption of fulvic acid depends on both the surface area of the adsorbent and solution pH. Fulvic acid was more susceptible to adsorption by SWCNTs and

MWCNTs than by activated carbon. The mechanisms of interplay between the apparent fulvic acid and the surface of CNT are quite complex and involve hydrogen bonding, electrostatic interaction, hydrophobic interaction (van der Waals forces) and π–π interactions. The adsorption of fulvic acid was reduced significantly with increasing pH due to increased electrostatic repulsion and the decrease of hydrophobic interactions and hydrogen bonds.

10.6 Concluding Remarks

As shown in this chapter, Nanotechnology can play an important role in this context and therefore an active area of emerging research is the development of new nanomaterials with elevated affinity, capacity and selectivity for adsorption. These nanomaterials, such as dendrimers, zeolites, nanocomposites based on metal hydrides and CNTs, have extremely interesting properties that relate to many applications. In fact, many potential applications of these nanostructured materials are being proposed, including obtaining devices for storage and energy conversion, sensors, nano-scale adsorbent devices, among others. Some of these applications are currently in the commercial phase, while others are still undergoing tests. In particular, the use of CNTs for hydrogen storage is still controversial [70].

Currently, the cost of these nanomaterials is highlighted as a major obstacle impeding its large-scale application in, for example, water treatment and energy storage. However, efficient and continuous production using low-cost resources will reduce, probably, the costs of production in the coming years [71].

References

[1] Savage N, Diallo MS (2005) Nanomaterials and water purification: opportunities and challenges. J Nanopart Res 7:331–342

[2] Savage N, Diallo M, Duncan J, Street A, Sustich R (2009) Nanotechnology Applications for clean water. Micro & nano technologies, William Andrew Inc., Norwich, NY

[3] Dillon AC, Jones KM, Bekkendahl TA, Kiang CH, Bethune DS, Heben MJ (1997) Storage of hydrogen in single-walled carbon nanotubes. Nature 386:377-379

[4] Ye Y, Ahn CC, Witham C, Fultz B, Liu J, Rinzler AG, Colbert D, Smith KA, Smalley RE (1999) Hydrogen adsorption and cohesive energy of single-walled carbon nanotubes. Appl Phys Lett 74:2307-2309

[5] Chen P, Wu X, Lin J, Tan KL (1999) High H_2 uptake by alkali-doped carbon nanotubes under ambient pressure and moderate temperatures. Science 285:91-93

[6] Lee SM, Park KS, Choi YC, Park YS, Bok JM, Bae DJ et al (2000) Hydrogen adsorption and storage in carbon nanotubes. Synth Met, 113:209-216

[7] Yildirim T, Ciraci T (2005) Titanium-decorated carbon nanotubes as a potential high-capacity hydrogen storage medium. Phys Rev Lett 94:175501

[8] Ioannatos GE, Verykios XE (2010) H_2 storage on single- and multi-walled carbon nanotubes. Int J Hydrogen Energy 35:622-628

[9] Sakintuna B, Lamari-Darkrim F, Hirscher M (2007) Metal hydride materials for solid hydrogen storage: A review. Int J Hydrogen Energy 32(9):1121-1140

[10] Lu CS, Chiu HS (2006) Adsorption of zinc(II) from water with purified carbon nanotubes. Chem Eng Sci 61:1138-1145
[11] Lu C, Liu C (2006) Removal of nickel(II) from aqueous solution by carbon nanotubes. J Chem Technol Biotechnol 81:1932–1940
[12] Li YH, Wang S, Luan Z, Ding J, Xu C, Wu D (2003) Adsorption of cadmium(II) from aqueous solution by surface oxidized carbon nanotubes. Carbon 41:1057–1062
[13] Li YH, Di Z, Ding J, Wu D, Luan Z, Zhu Y (2005) Adsorption thermodynamic, kinetic and desorption studies of Pb2+ on carbon nanotubes. Water Res 39:605–609
[14] Hu J, Chen C, Zhu X, Wanga X (2009) Removal of chromium from aqueous solution by using oxidized multiwalled carbon nanotubes. J Hazard Mater 162:1542–1550
[15] Pillaya K, Cukrowska EM, Coville NJ (2009) Multi-walled carbon nanotubes as adsorbents for the removal of parts per billion levels of hexavalent chromium from aqueous solution. J Hazard Mater 166:1067–1075
[16] Kuo CY (2009) Water purification of removal aqueous copper(II) by as-grown and modified multi-walled carbon nanotubes. Desalination 249:781–785
[17] Moreno N, Querol X, Ayora C, Pereira CF, Janssen-Jurkovicová M (2001) Utilization of zeolites synthesized from coal fly ash for the purification of acid mine waters. Environ Sci Technol 35:3526-3534
[18] Kuo CY, Wu CH, Wuc JY (2008) Adsorption of direct dyes from aqueous solutions by carbon nanotubes: determination of equilibrium, kinetics and thermodynamics parameters. J Colloid Interface Sci 327:308–315
[19] Yao Y, Xu F, Chen M, Xu Z, Zhu Z (2010) Adsorption behavior of methylene blue on carbon nanotubes. Bioresour Technol 101:3040–3046
[20] Wu CH (2007) Adsorption of reactive dye onto carbon nanotubes: equilibrium, kinetics and thermodynamics. J Hazard Mater 144:93-100
[21] Mishra AK, Arockiadoss T, Ramaprabhu S (2010) Study of removal of azo dye by functionalized multi walled carbon nanotubes. Chem Eng J 162:1026-1034
[22] Upadhyayula VKK, Deng S, Mitchell MC, Smith GB (2009) Application of carbon nanotube technology for removal of contaminants in drinking water: A review. Sci Total Environ 408:1–13
[23] Upadhyayula VKK, Deng S, Smith GB, Mitchell MC (2009) Adsorption of *Bacillus subtilis* on single walled carbon nanotube aggregates, activated carbon and nanoceramTM. Water Res 43:1–9
[24] Yang K, Xing B (2009) Adsorption of fulvic acid by carbon nanotubes from water. Environ Pollut 157:1095-1100
[25] Brunauer S (1945) The adsorption of gases and vapors. vol.1. Princeton University Press, London
[26] De Boer JH (1968) The dynamical character of adsorption. 2nd Ed. Clarendon Press, Oxford
[27] Dąbrowski A (2001) Adsorption - from theory to practice. Adv Colloid Interface Sci 93:135-224
[28] Tomalia DA (2005) Birth of a new macromolecular architecture: dendrimers as quantized building blocks for nanoscale synthetic polymer chemistry. Prog Polym Sci 30:294-324
[29] De Oliveira JMA (2009) New nanotechnology approaches using dendrimers modified with natural polymers for controlling stem cells behaviour in tissue engineering strategies. Thesis, University of Minho

[30] D'Emanuele A, Jevprasesphant R, Penny J, Attwood D (2004) The use of a dendrimer-propranolol prodrug to bypass efflux transporters and enhance oral bioavailability. J Controlled Release 95:447-453
[31] Fréchet JMJ, Tomalia DA (2001) Dendrimers and other dendritic polymers. Wiley and Sons, New York
[32] Diallo MS, Christie S, Swaminathan P, Johnson Jr JH, Goddard III WA (2005) Dendrimer enhanced ultrafiltration. 1. Recovery of Cu(II) from aqueous solutions using PAMAM dendrimers with ethylene diamine core and terminal NH_2 groups. Environ Sci Technol 39:1366-1377
[33] Smith JV (1984) Definition of a zeolite. Zeolites 4:309-310
[34] Camblor MA, Corma A, Valencia S (1998) Characterization of nanocrystalline zeolite Beta. Microporous Mesoporous Mater 25(1-3):59-74
[35] Ding L, Zheng Y (2007) Nanocrystalline zeolite beta: The effect of template agent on crystal size. Mater Res Bull 42:584-590
[36] Song W, Grassian VH, Larsen SC (2005) High yield method for nanocrystalline zeolite synthesis. Chem Commun 36:2951–2953
[37] Tavolaro A, Tavolaro P, Drioli E (2007) Zeolite inorganic supports for BSA immobilization: Comparative study of several zeolite crystals and composite membranes. Colloids Surf B 55:67–76
[38] Song W, Justice RE, Jones CA, Grassian VH, Larsen SC (2004) Synthesis, characterization, and adsorption properties of nanocrystalline ZSM-5. Langmuir 20: 8301-8306
[39] Song W, Li G, Grassian VH, Larsen SC (2005) development of improved materials for environmental applications: Nanocrystalline NaY zeolites. Environ Sci Technol 39:1214-1220
[40] Iijima S, Ichihashi T, (1993) Single-Shell Carbon Nanotubes of 1-nm Diameter. Nature 363:603-605.
[41] Bethune DS, Kiang CH, De Vries MS, Gorman G, Savoy R, Vazques J, Beyers R, (1993) Cobalt-catalysed growth of carbon nanotubes with single-atomic-layer walls. Nature 363:605-607
[42] Iijima S (1991) Helical microtubules of graphitic carbon. Nature 354:56-58
[43] Klumpp C, Kostarelos K, Prato M, Bianco A (2006) Functionalized carbon nanotubes as emerging nanovectors for the delivery of therapeutics. Biochim Biophys Acta 1758:404-412
[44] Banerjee S, Hemraj-Benny T, Wong SS (2005) Covalent surface chemistry of single-walled carbon nanotubes. Adv Mater 17:17-29
[45] Serp P, Corrias M, Kalck P (2003) Carbon nanotubes and nanofibers in catalysis. Appl Catal A Gen 253:337-358
[46] Yang C-M, Kaneko K, Yudasaka M, Iijima S (2002) Surface chemistry and pore structure of purified HiPco single-walled carbon nanotube aggregates. Physica B 323:140–142
[47] Niu JJ, Wang JN, Jiang Y, Su LF, Ma J (2007) An approach to carbon nanotubes with high surface area and large pore volume. Microporous Mesoporous Mater 100:1-5
[48] Li F, Wang Y, Wang D, Wei F (2004) Characterization of single-wall carbon nanotubes by N_2 adsorption. Carbon 42:2375-2383
[49] Yang QH, Hou PX, Bai S, Wang MZ, Cheng HM (2001) Adsorption and capillarity of nitrogen in aggregated multi-walled carbon nanotubes. Chem Phys Lett 345:18-24

[50] Babaa MR, Dupont-Pavlovsky N, McRae E, Masenelli-Varlot K (2004) Physical adsorption of carbon tetrachloride on as-produced and on mechanically opened single walled carbon nanotubes. Carbon 42:1549-1554
[51] Agnihotri S, Mota JPB, Rostam-Abadi M, Rood MJ (2006) Adsorption site analysis of impurity embedded single-walled carbon nanotube bundles. Carbon 44:2376-2383
[52] Cheng J, Yuan X, Zhao L, Huang D, Zhao M, Dai L, Ding R (2004) GCMC simulation of hydrogen physisorption on carbon nanotubes and nanotube arrays. Carbon 42:2019-2024
[53] Zubizarreta L, Arenillas A, Pis JJ (2009) Carbon materials for H_2 storage. Int J Hydrogen Energy 34:4575-4458
[54] Orimo S, Züttel A et al (2003) Hydrogen interaction with carbon nanostructures: current situation and future prospects. J Alloys Compd 356–357:716-719
[55] Panella B, Hirscher M, Roth S (2005) Hydrogen adsorption in different carbon nanostructures. Carbon 43:2209–2214
[56] Liu F, Zhang X, Cheng J, Tu J, Kong F, Huang W, Chen C (2003) Preparation of short carbon nanotubes by mechanical ball milling and their hydrogen adsorption behavior. Carbon 41:2527-2532
[57] Darkrim FL, Levesque D (1998) Monte Carlo simulations of hydrogen adsorption in single-walled carbon nanotubes. J Chem Phys 109:4981-4984
[58] Gülseren O, Yildirim T, Ciraci S (2002) Effects of hydrogen adsorption on single-wall carbon nanotubes: Metallic hydrogen decoration. Phys Rev B 66:121401-1-121401-4
[59] Schlapbach L, Züttel A (2001) Hydrogen storage-materials for mobile applications. Nature 414:23-31
[60] Züttel A, Wenger P, Sudan P, Mauron P, Orimo S-I (2004) Hydrogen density in nanostructured carbon, metals and complex materials. Mater Sci Eng B 108:9–18
[61] Lima GF, Peres MM, Garroni S et al (2010) Microstructural characterization and hydrogenation study of extruded MgFe alloy. J Alloys Compd 504:S299-S301
[62] Huot J, Liang G, Boily S, van Neste A, Schulz R (1999) Structural study and hydrogen sorption kinetics of ball-milled magnesium hydride. J Alloys Compd 293-295: 495-500
[63] Diallo MS (2009) Water treatment by dendrimer-enhanced filtration: Principles and applications. In S. Nora, D. Mamadou, D. Jeremiah, S. Anita & S. Richard (Eds.), Nanotechnology Applications for Clean Water. William Andrew Publishing, Boston
[64] Diallo MS, Christie S, Swaminathan P, Balogh L, Shi X, Um W, Papelis C, Goddard III WA, Johnson Jr JH (2004) Dendritic chelating agents 1. Cu(II) binding to ethylene diamine core poly(amidoamine) dendrimers in aqueous solutions. Langmuir 20: 2640-2651
[65] Álvarez-Ayuso E, García-Sánchez A, Querol X (2003) Purification of metal electroplating waste waters using zeolites. Water Res 37:4855-4862
[66] Yantasee W, Lin Y, Fryxell GE, Busche BJ, Birnbaum JC (2003) Removal of heavy metals from aqueous solution using novel nanoengineered sorbents: Self-assembled carbamoylphosphonic acids on mesoporous silica. Separ Sci Technol 38:3809–3825
[67] Kelly SD, Kemmer KM, Fryxell GE, Liu J, Mattigod SV, Ferris KF (2001) X-ray-absorption fine-structure spectroscopy study of the interactions between contaminant tetrahedral anions and self-assembled monolayers on mesoporous supports. J Phys Chem B 105:6337-6346
[68] Long RQ, Yang RT (2001) Carbon nanotubes as superior sorbent for dioxin removal. J Am Chem Soc 123:2058-2059

[69] Rao GP, Lu C, Su F (2007) Sorption of divalent metal ions from aqueous solution by carbon nanotubes: A review. Sep Purif Technol 58:224-231

[70] Kajiura H, Tsutsui S, Kadono K, Kakuta M, Ata M, Murakami Y, (2003) Hydrogen storage capacity of commercially available carbon materials at room temperature. Appl Phys Lett 82:1105-1107

[71] Agboola AE, Pike RW, Hertwig TA, Lou HH (2007) Conceptual design of carbon nanotube processes. Clean Technol Environ Policy 9:289-311

Abbreviations

CNF - carbon nanofibers
CNT(s) – carbon nanotube(s)
DEUF - dendrimer-enhanced ultrafiltration
MWCO - molecular weight cut-off
MWCNT(s) – multi-walled carbon nanotube(s)
PAMAM - poly(amidoamine)
SAMMS - Self-assembled monolayers on mesoporous supports
SEM - scanning electron microscopy
SWCNT(s) – single-walled carbon nanotube(s)
TEM - transmission electron microscopy
UF - ultrafiltration

11 Use of Natural and Modified Natural Nanostructured Materials

André Zimmer[1], Mônica Jung de Andrade[2], Felipe Antonio Lucca Sánchez[2], and Antonio Shigueaki Takimi[2]

[1] Instituto Federal de Educação, Ciência e Tecnologia do Rio Grande do Sul, 95770000, Feliz, Brazil
[2] Departamento de Engenharia de Materiais, Universidade Federal do Rio Grande do Sul, 90035190, Porto Alegre, Brazil
E-mail: mja0612@gmail.com

Abstract. The use of natural nanomaterials is environmentally friendly, socially responsible and cost effective. Hence, several industries are investing in cheap ways to explore and process natural resources to make natural nanomaterials available. Natural nanomaterials are abundant; however, certain drawbacks such as incompatibility and extraction are a challenge to make their use more advantageous. Montmorillonite is the most common natural nanomineral used in industry and nanocellulose could be extracted from the most important skeletal component in plants, cellulose. In this chapter a brief overview is given of natural and modified natural nanostructured materials, with emphasis on layered silicates (clays) and cellulose-based materials, followed by some industrial examples.

Keywords: clays, cellulose, natural stamps, halloysite nanotubes, amyloids.

11.1 Introduction

There are many natural (animal, vegetal and mineral) nanomaterials around us and their properties are directly connected to their molecular supra-organization.

Natural nanomaterials are materials with at least one dimension in nanoscale that is a result of natural processes. Some particles arising from aerosols of marine, mineral, volcanic, biogenic and cosmic origins could be considered natural nanomaterials. Aerosols originated from natural sources can be directly emitted as particles (primary aerosols) or they can also be the result of chemical reactions (secondary aerosols).

Natural and modified natural nanostructured materials can have many advantages in comparison to synthetic nanomaterials, like having fewer safety concerns, being socially responsible, combining functionality with an eco-friendly approach and, in certain cases, demonstrating economic advantages. However, certain drawbacks such as incompatibility with the hydrophobic polymer matrix, the tendency to form aggregates during processing, and poor resistance to moisture greatly reduce the potential of natural fibers to be used as reinforcement in polymers [1].

The top–down approach, which starts with the modification (like breaking or separating) of large structures and proceeds to smaller structures, has been widely used to create nanostructured materials for various industrial applications. In order to prepare these nanostructured materials (like nanoparticles and nanocomposites) there are many methods such as ball milling, ultrasonication, reverse micelles, chemical reduction, chemical and physical vapor depositions, solid state reaction, hydrothermal and microwave. Their applications can be as wide as refractories, textiles, energy, biomedicals, functional barriers and environmental fields [2,3].

According to Gao et al. [4], natural materials such as bone, tooth, and nacre are nanocomposites of proteins and minerals with superior strength. According to Doktycz and Simpson [5], these biological system assembly efforts are particularly instructive, as we gain command over the directed synthesis and assembly of synthetic nanoscale structures. This field is most likely to advance through the transfer of biological principles of organization into the bottom–up synthesis of complex synthetic nanoscale material systems. Attention to principles of systems biology can lead to the practical development of nanoscale technologies with possible realization of synthetic systems with cell-like complexity.

Gao et al. [4] show that the nanocomposites in nature exhibit a generic mechanical structure in which the nanometer size of mineral particles is selected to ensure optimum strength and maximum tolerance of flaws (robustness). They further show that the widely used engineering concept of stress concentration at flaws is no longer valid for nanomaterial design.

11.2 Natural Layered Silicates – Bentonite Clays

Clays are an earthy and naturally occurring material with a fine grain size and certain plasticity when mixed with a suitable liquid, like water. Clays are essentially composed of crystalline particles from a small group of minerals, referred to as the clay minerals, which are a class of hydrated phyllosilicates making up the fine-grained fraction of rocks, sediments, and soils. Chemically, clays are formed by the mixture of aluminum and iron silicates, further containing, usually, certain amounts of alkaline and alkaline-earthy elements [6,7].

Because the group of clay minerals is vast, they can be classified by different aspects of interest like: chemical composition, interlayer material content and morphological structure. The *Association Internationale pour l'Etude des Argiles* recommends the following sub-division of clay minerals in two general classes [8]:

1) crystalline silicates with layered structure;
2) crystalline silicates with fibrous structure.

In industrial applications of clays one distinguishes four types of clays [7]:

1) bentonites, with montmorillonite as the principal clay mineral constituent;
2) kaolins containing kaolinite;
3) palygorskite and sepiolite;
4) 'common clays' which often contain illite/smectite mixed-layer minerals, and are largely used for traditional ceramics manufacture.

Montmorillonite is the most common clay mineral used in polymer-matrix clay reinforced nanocomposites (PCN) and is composed of an octahedral alumina sheet sandwiched between two tetrahedral silica sheets. The anionically charged sheets, or clay platelets, are strongly held together by cations, such as Na^+, Li^+, Ca^{2+}, Fe^{2+} and Mg^{2+}. The thickness of a single clay sheet is about 1 nm. The tightly bound stacks of clay sheets may be separated by a simple dispersion in a polar solvent, such as water. Natural clay is also miscible with polar polymers in which the platelets are readily dispersed. However, the natural separation of clay sheets in non-polar polymer such as polyolefins is quite difficult [7].

In general, bentonite is any clay composed predominantly of montmorillonite clay mineral of the smectite group whose main properties are: particles of colloidal size, high degree of layer stacking disorder, high specific surface area, moderate layer charge, large cation exchange capacity, variable interlayer separation, depending on ambient humidity, propensity for intercalating extraneous substances, including organic compounds and macromolecules, and ability of some members (e.g., Li^+ and Na^+ exchanged forms) to show extensive interlayer swelling in water; under optimum conditions, the layers can completely dissociate. It is also referred to as exfoliated clay [7].

In the development of new nanotechnological materials, bentonite clays have been extensively employed as a "nanofiller" in several polymeric matrixes and it has been demonstrated that by using small amounts of it (less than 5 wt. %), it is possible to increase polymer properties regarding the mechanical and thermal resistance and, mainly, the flame retardant and gas barrier behavior [9,10,11,12,13,14,15,16,17,18].

Three main types of (nano)composites can be obtained when layered clays like bentonites are associated with polymers. When the polymer chains cannot be mixed well among the clay layers, the structure obtained shows tactoids dispersed with immiscibility in the polymer matrix and with properties that do not present relevant improvements in the material, which is normally classified as a conventional composite or microcomposite. On the other hand PCN can be obtained when the polymer chains are intercalated between swollen clay layers with their regular alteration of galleries and laminates (intercalated PCN) or when the layered clay is totally delaminated/exfoliated and dispersed in the polymer matrix (exfoliated PCN). Intercalated PCN has a layer distance of a few nanometers; that is, the space occupied by the polymer chain and exfoliated PCN loses its ordered structure and the distance between the layers is about the size of the gyration radius of the polymer. Both classes of PCN are schematically compared with an immiscible system, denominated as a microcomposite, as shown in Figure 11.1. It can be said that an intercalated PCN displays limited miscibility and that exfoliated PCN is totally miscible. Both of these hybrid structures can co-exist in the polymer matrix [7].

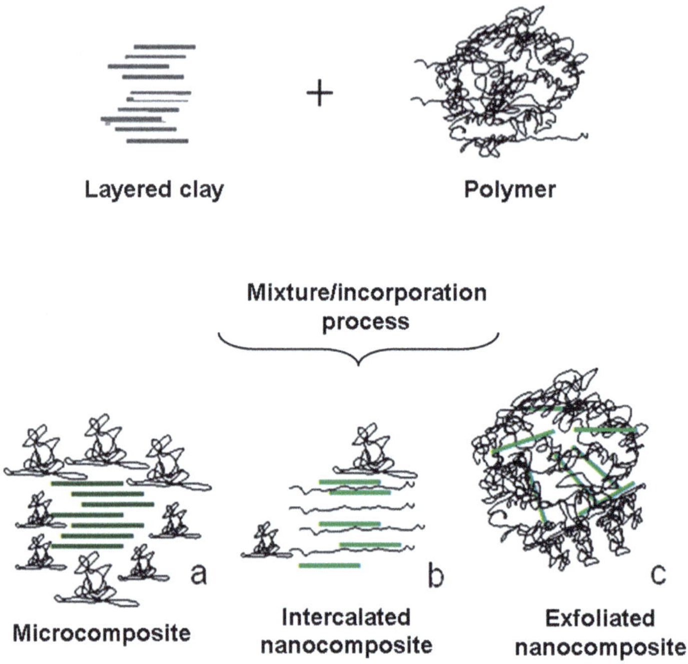

Fig. 11.1. Schematic representation of composites and nanocomposites obtained by the use of layered clays.

The mixing of a typical petroleum-based polymer and natural bentonite leads to the formation of a microcomposite, as this process leads to the dispersion of micron-size stacked sheets. This behavior is due to the poor physical interaction of the polar surfaces of layered silicates with the apolar macromolecules. If these interactions become stronger, then the layered silicates can be dispersed in the polymer matrix at nanometer scale. The layered silicates like montmorillonite present a very high aspect ratio (10–1000) and a large surface area (above 700 m^2/g). Hence, for PCN containing low concentrations of layered silicates, the total interface between polymer and layered silicates is much greater than that of conventional composites for the same concentration [19,20,21].

The layered silicates are only compatible with water soluble polymers, such as poly(vinyl alcohol) [22]. To satisfy wetting criteria, layered silicates like bentonite are exfoliated in hot water under high shear stress, and then functionalized by ion exchange reactions with cationic surfactants including quaternary alkylammonium cations or phosphonium cations, where at least one of these substituents must be a long carbon chain (12-18 carbon atoms) in order to make the silicate layer compatible with the polymer matrix [23]. This material is commonly named organoclay or nanoclay.

The surfactant molecules attached to the surface of the silicate layer lowers the surface energy, improves the wetting characteristics of the silicate and causes an increase in the basal spacing d(001), enabling the macromolecules to enter the interlayer spacing. In some cases, the alkylammonium cations could provide functional groups that can react with the polymer or initiate polymerization of monomers to improve the strength of the interface between the inorganic and the polymer, as developed by Toyota researchers in the mid-80s [21,24].

For complete exfoliation to occur, the interfacial tension between the organoclay and the polymer matrix must be low enough to permit particle wetting. However, this condition is not expected to be sufficient to promote exfoliation, and a high motional disorder of the surfactant chains is also required [25].

PCN and other polymer-layered silicates are typically synthesized by three methods referred as:

(1) In-situ polymerization method:
In this process, mechanical mixing of a clay mineral in a monomer solution is required. With an appropriated mixing system the monomer is intercalated within the interlayer of the clay, promoting a delamination. The polymerization initiated by a number of ways follows to yield linear or cross-linked polymer matrices. In some cases it is necessary that clays be pretreated by a pre-intercalation step of long chain alkyl ammonium ion intercalation to aid exfoliation [19-21].

(2) Solution method:
The intercalation of the polymer from the solution involves a solvent system in which polymers are soluble and the silicate layers are swellable. When the polymer and a layered clay solution are mixed, the polymer chains dissolved in the solvent intercalate into the interlayer of the clays with the solvent. An intercalated structure remains, resulting in PCN after the solvent evaporation occur [19-21].

(3) Direct melt mixing method:
As indicated by the name, in this process a direct intercalation of the molten polymer into the layered clays takes place. It is a method where solvents are not used. The polymer is heated above the glass transition temperature in either static or flow condition and mixed, normally in a twin screw extrusion system, with the layered silicates directly. Modified layered clays are usually employed to promote intercalation. The polymer chains from the molten mass spread into the silicate galleries to form either intercalated or exfoliated PCNs according to the degree of penetration [19-21].

Polymer nanocomposites enable substantial improvements in material properties such as shear and bulk modulus, yield strength, toughness, scratch resistance, optical properties, electrical conductivity, gas and solvent transport, with a small loading of clay dispersed in the polymer matrix [19]. There are several excellent reviews about the properties of polymer nanocomposites [19,23,26,27,28,29,30,31], to name a few. For example, the addition of only 2%wt of clay in Nylon-6 improves strength, modulus, and, in particular, heat distortion temperature (HDT); and these properties are saturated at just 5 wt.% of clay [21]. In olefinic polymers like polypropylene, the Young modulus, tensile and impact strength, flammability and HDT are improved with the addition of organoclay [26]. Also, thermoset polymers and elastomer can be reinforced using small amounts of organoclays.

Because of its morphology, montmorillonites will develop similar increases in modulus and tensile strength at a 3-5 wt.% loading compared to 20-60 wt.% loading of conventional reinforcing agents such as kaolin, silica, talc, and carbon black. Implicit advantages include lighter plastic parts with greater transparency.

Its particle shape also benefits polymers in terms of an increase of barrier properties to moisture, solvents, chemical vapors, gases such as O_2 and flavors. It is this morphology that leads to the improved permeation barrier through a tortuous path mechanism. With montmorillonites, polymers will have increased dimensional stability at low reinforcement loadings. Dramatic decreases in the coefficient of linear thermal expansion values have been reported.

Polymers can achieve a higher HDT, with a load of a few percent of montmorillonite. These nanofillers will increase the temperature at which the polymer will start to acquire a softened state. This property is critical, for example, in under-hood automotive applications.

Also important is that the thermoplastic polymer will be more recyclable. Montmorillonite performance actually improves upon recycling.

Moreover, PCN offers a synergistic flame-retardant approach. The improved flame retardancy, as measured by Cone Calorimetry, shows a decrease in the Peak Heat Release Rate. A decrease in smoke and an increase in char formation are observed. Combination with traditional flame retardants can enable the passing of specified flame tests.

Furthermore, the polymer also will be easily colored due to the colloidal nature, high surface area, and surface treatability of montmorillonite (it can serve as an active site to fix dyes into polymers). The appearance of painted parts is improved compared to conventional reinforced parts. The nanocomposite particles are much smaller than traditional reinforcing agents so the polymer surface is much smoother. There is even a reduced static cling in films containing nanocomposites. This property was observed in some work being done by Southern Clay Products personnel.

11.3 Cellulose-Based Nanostructured Materials

As the most important skeletal component in plants, polysaccharide cellulose is an almost inexhaustible polymeric raw material with fascinating structure and properties. Formed by the repeated connection of d-glucose building blocks, the highly functionalized, linear stiff-chain homopolymer is characterized by its hydrophilicity, chirality, biodegradability, broad chemical-modifying capacity, and its formation of versatile semicrystalline fiber morphologies [32].

The supramolecular structure of cellulose includes amorphous and crystalline regions. The crystalline regions are up to several microns long and a few nanometers wide. Because of their dimensions, i.e., their high aspect ratio they can be regarded as nano-whiskers. It is further possible to isolate these nano-crystals in the form of independent particles. The size of the cellulose nano-whiskers is controlled by the size of the crystalline regions in the cellulose and therefore by the biomass sources.

Development of nanocomposites based on nanocellulosic materials is a rather new, but rapidly evolving research area. Cellulose is abundant in nature, biodegradable and relatively cheap, and it is a promising nano-scale reinforcement material for polymers [33].

Based on a review by Siró and Plackett [33], nano-scale cellulose fiber materials (e.g., microfibrillated cellulose and bacterial cellulose) serve as promising candidates for bio-nanocomposite production since cellulose is abundant, has high strength and stiffness, low weight, and biodegradability.

According to Nakagaito and Yano [34] it is possible to manufacture a composite of high-strength using microfibrillated cellulose (MFC) derived from kraft pulp. Because of the unique structure of nano-order-scale interconnected fibrils and microfibrils greatly expanded in the surface area that characterizes MFC, it was possible to produce composites that exploit the extremely high strength of microfibrils. The Young modulus (E) and bending strength (σ_b) of composites using phenolic resin as binder achieved values up to 19 GPa and 370 MPa, respectively, with a density of 1.45 g/cm^2, exhibiting outstanding mechanical properties for a plant-fiber-based composite.

Nakagaito, Iwamoto and Yano [35] produced high-strength composites using bacterial cellulose (BC) sheets impregnated with phenolic resin and compressed at 100 MPa. By utilizing this unique material synthesized by bacteria, it was possible to improve the mechanical properties over the previously reported high-strength composites based on fibrillated kraft pulp of plant origin. BC-based composites were stronger, and in particular the Young modulus was significantly higher, attaining 28 GPa versus 19 GPa of fibrillated pulp composites. The superior modulus value was attributed to the uniform, continuous, and straight nano-scalar network of cellulosic elements oriented in-plane via the compression of BC pellicles. With a sulfuric acid concentration of 63.5% (w/w), it was possible to produce cellulose whiskers with a length between 200 and 400 nm and a width less than 10 nm in approximately 2 h and with a yield of 30% (of initial weight). Disintegration of amorphous regions and degradation of crystalline parts during hydrolysis is probably the explanation for the low yield. The whiskers produced under these treatment conditions would be well suited for incorporation into a polymer using an extrusion process where nanobiocomposites are produced.

Cellulose is hydrophilic, naturally abundant and finds a great number of applications. Cellulose is a polymer naturally synthesized by plants and the most important component of their cell walls. Cellulose owes its importance to the mechanical strength of its fibrils (in particular due to the high level of intermolecular hydrogen bonding), as well as to its chemical properties that allows for a great amount of chemicals to derive from it.

The use of agricultural residues for isolation of cellulose nanofibers and subsequently making nanocomposites is a prospective commercial application that would unlock the potential of these underutilized renewable materials and provide a non-food based market for the agricultural industry.

An example to obtain nanocellulose is the method of Acid Hydrolysis that follows this sequence: (i) Bleached softwood kraft pulp; (ii) 8% Cellulose + 64% H_2SO_4; (iii) Stirred at 45 °C, 45 min to ~ 1 hour; (iv) Diluted with water; (v) Centrifuged, washed and neutralized; (vi) Re-dispersed with ultrasonication; (vii) Allowed to stand over a mixed bed resin for 48 h; (viii) The mixture is centrifuged and the supernatant is filtered through filter paper. The filtrate is a colloidal nanowhisker suspension. With a sulfuric acid concentration of 63.5% (w/w), it is possible to produce cellulose whiskers with a length between 200 and 400 nm and a width of less than 10 nm in approximately 2 h and with a yield of around 30% (of initial weight). Disintegration of amorphous regions and degradation of crystalline parts during hydrolysis is probably the explanation for the low yield. The whiskers produced under these treatment conditions would be well suited for incorporation into a polymer using an extrusion process where nanobiocomposites are produced.

11.4 Others

11.4.1 Natural Stamps

In the work of Zhang et al. [36], the direct use of natural nanostructures as stamps for nanoimprint lithography is reported. With these natural stamps, nanowell arrays (negative structures of cicada wings) have been fabricated on a polymer support. Furthermore, the nanowell arrays can be transferred to the silicon substrate by reactive ion etching, and thus exhibit an antireflective property. With patterned poly(methyl methacrylate) as a mold, hexagonal gold-pillar arrays similar to the surface of cicada (*Cryptympana atrata Fabricius*) wings can also be obtained by thermodeposition.

As shown in Figure 11.2, the microscopic structures of the cicada wings consist of ordered hexagonal close-packed arrays of pillars with a spacing of about 190 nm [36].

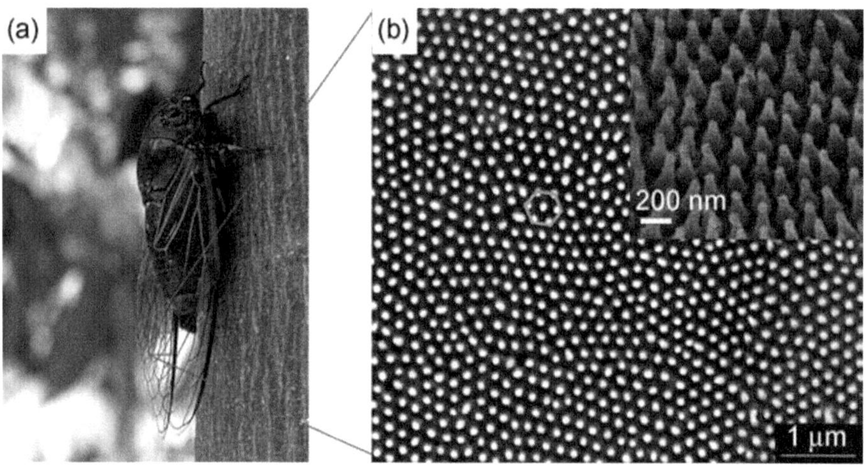

Fig. 11.2. (a) Photograph of *Cryptympana atrata Fabricius*. (b) Scanning electron microscopy (SEM) image of the surface of cicada wings. The space between each pillar is about 190 nm. The pillars are arranged in a hexagonal array. Inset: SEM image showing a tilt of about 30° [36]. (Copyright Wiley-VCH Verlag GmbH & Co. KGaA. Reproduced with permission.)

11.4.2 Halloysite Nanotubes

Naturally formed in the earth over millions of years, halloysite nanotubes are unique and versatile nanomaterials that are formed by surface weathering of aluminosilicate minerals and are composed of aluminum, silicon, hydrogen and oxygen [37]. Halloysite nanotubes are ultra-tiny hollow tubes with diameters typically smaller than 100 nanometers (range from about 40 nm to 200 nm), with lengths typically ranging from about 500 nm to over 1.2 μm.

The functional characteristics desired for specific applications can be controlled through selection of nanotube diameter and length. Furthermore, halloysite nanotubes can be coated with metallic and other substances to achieve a wide variety of electrical, chemical, and physical properties. Their applications include additives in polymers and plastics, electronic components, cosmetics, and home and personal care products. For example, they can be filled with such things as active ingredients including many that are used for cosmetics, household and personal care products, pesticides, pest repellents, pharmaceuticals and other agents that could benefit from extended release [37].

11.4.3 Amyloids

Amyloids are insoluble fibrous protein aggregates sharing specific structural traits. Amyloid fibers constitute one of the most abundant and important naturally occurring self-associated assemblies. A variety of protein and peptide molecules with various amino acid sequences form these highly stable and well-organized assemblies under diverse conditions. These assemblies display phase states ranging from liquid crystals to rigid nanotubes. A range of potential technological applications rely on amyloid-fiber-based materials. According to Cherny and Gazit [38], amyloids are excellent candidates for the fabrication of molecular nanobiomaterials (e.g., wires, layers, gels, scaffolds, templates, and liquid crystals) using the "bottom-up" strategy - method to produce nanoparticles from atoms or molecules. This is because of their structural compatibility, nanoscale dimensions, efficient assembly into well-defined ultrastructures, ease of production, and low cost. Furthermore, the building blocks may be varied extensively by rather simple protein-engineering techniques [38].

11.5 Industrial Examples of the Use of Modified and Non-modified Natural Nanostructured Materials

Nanocellulose, nanostructured minerals (like clay halloysite nanotubes, modified bentonites and montmorillonites), organic–inorganic hybrid nanomaterials and PCN are some of the examples of nanostructured natural materials that are being commercialized and used by different companies around the world, as will be shown below [39-53].

Nanocellulose, for example, is described as being as strong and light as Kevlar, environmentally compatible and renewable. Thus, nanocellulose applications include [39,40]: moldable light weight and high strength materials, medical implants, electrodes for fuel cells, barrier film for packaging applications (keeping oxygen from spoiling food), composites for construction, vehicles and furniture. For example, Sony Corporation, in conjunction with Ajinomoto and the Japanese Agency of Industrial Science and Technology developed the first audio speaker diaphragms using microbial cellulose [41]. The unique dimensional stability of microbial cellulose gives rise to a sound transducing membrane which maintains high sonic velocity over a wide range of frequencies, therefore being the best material to meet the rigid requirements for optimal sound transduction. In the near future it is expected that larger speaker diaphragms will be made of microbial cellulose. Sony Corporation and TOPPAN Printing Co., announced the successful development of 25 GB paper disc based on Blue-ray Disc technology, the recording layer on which the data is stored lies under a protective layer [42].

Recently, the French Imerys [43], global leader in the supply of high-quality white pigments for the world's paper industry, also patented spherical and rod-shaped proppants and anti-flowback agents based on nanostructured minerals from the group consisting of kyanite, sillimanite, and andalusite (may be used alone or in combination with other materials, such as bauxite, kaolin, meta-kaolin, pure or technical-grade alumina) which also possess high strength and high conductivity.

Studies have shown that organic–inorganic hybrid nanostructured materials not only increase mechanical, thermal and gas-barrier properties of polymers, but also are useful for pollution control and water treatment [3,44]. Indeed, the growing market of nanostructured natural inorganic materials for nanocomposite applications can be visualized through the many achievements (patents and products) of companies like NaturalNano Inc. [45], Nanocor [47] and Nanoclay [48], for example.

NaturalNano Inc. [45] is a nanomaterials company developing technologies and processes with a focus on the fast-growing market of PCN (mostly with halloysite and kaolin clays) for a wide range of applications: automotive (lighter parts to allow for increased transportation loads), military (make vehicles lighter, so they will be cheaper to transport), packaging (stronger or thinner, e.g., storage bags, food packaging), aerospace, and electronics applications. The company holds over twenty issued or pending patents and proprietary know-how for extraction and separation processes, of halloysite and other nanotubes, in combination with other materials. For example, Pleximer™ is a PCN with clay halloysite nanotubes, a naturally occurring two-layered aluminosilicate, non-toxic and environmentally-friendly nano material. According to them, future opportunities lie in the ability to fill halloysite nanotubes with active ingredients that will allow its use in cosmetics, odor masking, agriculture, medicine, and many other applications requiring extended release patents available for licensing.

Sigma-Aldrich [46] also commercializes some natural nanomaterials like halloysite nanotubes and montmorillonite nanoclays for polymer composites. Their halloysite nanotubes have dimensions varying within 1-15 μm of length and 10-150 nm of inner diameter (depending on the deposits) and their product applications include: (i) controlled release of anticorrosion agents; (ii) sustained release of herbicides, insecticides, fungicides and anti-microbials; (iii) sustained release of drugs, food additives, and fragrances; (iv) templating synthesis of rod-like nanoparticles; (v) uses as catalytic supports and molecular sieves; (vi) specific ion adsorption; (vii) use as plastic fillers for strength reinforcement and scratch protection; (viii) use in advanced ceramic materials, especially biocompatible implants. Polymer-clay nanocomposites based on plate-like montmorillonite (which consists of ~ 1 nm thick aluminosilicate layers surface-substituted with metal cations and stacked in ~ 10 µm-sized multilayer stacks) can increase mechanical strength, decrease gas permeability, superior flame-resistance, and even enhance transparency (dispersed nanoclay plates suppress polymer crystallization) compared to the pure polymer.

Nanocor [47] is a new operating subsidiary of AMCOL International Corporation and is the largest global supplier of Nanomer® clays (registered trademark of Nanocor Corp) specifically designed for PCN. Some of their clay products are modified bentonite and montmorillonite. Their product can speed up cure time, improve processing, decrease gas permeability, improve modulus, enhance chemical resistance, improve flame retardancy and increase the HDT when incorporated to different polymers when compared to pure polymers. To enhance the compatibility with hydrophobic polymers, the mineral's surface is exchanged with alkylammonium cations, for example. Some of their commercialized products can be used to reinforce polymers like epoxy, polyurethane, polypropylene, polyethylene, ethylene vinyl acetate, polyamides and hydrophilic polymers like polyvinyl alcohol).

Nanoclay fabricates Cloisite® additives (Figure 12.3), which consist of organically modified nanometer scale, layered magnesium aluminum silicate platelets. The silicate platelets that Cloisite® additives are derived from, are 1 nm thick and 70 – 150 nm across [48].

Fig. 11.3. Micrograph of Cloisite.

The North-American AlliedSignal Inc. (aerospace, automotive and engineering company) patented a nanocomposite material comprising a polymer as matrix having dispersed therein platelet clay mineral particles of montmorillonite that act as nanoscale barrier layers, which impart lower permeability to polymers than do comparable loadings of conventional barrier fillers [49]. Ford also patented similar polymeric thermoplastic nanocomposites reinforced by layered silicate clay (montmorillonite or kaolin) and a second filler material selected from the group consisting of calcium carbonate, titanium dioxide, talc, zirconium dioxide, zinc oxide, calcium silicate, aluminum silicate, calcium sulfate, alumina trihydrate, glass fibers, carbon fibers, and mixtures thereof [50].

Rheox, Inc., a leader in rheological additive products, patented hybrid organoclay (smectite clays) chemical compositions useful as additives to polymer, plastic and resin matrices to produce nanocomposites and nanocomposites containing such compositions [51]. One year later, Elementis Specialties Inc., another North-American chemicals manufacturing company, specialized in chromium and rheological additives, got the patent for a thermoplastic polymer blended with organoclay composition containing smectite [52] that resulted in nanocomposites with enhanced structural and flame-retardant properties [53].

This trend is also observed in final users, especially in the automotive industry. Some examples of automotive companies that are already making use of natural modified nanostructured materials are: the Italian Fiat [54], the German Mercedes Benz [55], the German Volkswagen [56], the French PSA Peugeot Citröen [57], the North-American Ford Motor Company [58], the North-American General Motors Company [59], the Japanese Mitsubishi Motors Corp. [60] and the Japanese Toyota Motor Corp. [61]. These companies are using from wood powder to vegetal fibers (like *Ananas erectifolius*, *Cocos nucifera* L., *Agave sp* and *Corchorus sp*) in components of the seats or doors inserts or panels or under-hoods or car trunks. One example is the Mercedes-Benz A-Class, which contains 27 components containing vegetal fibers [55].

11.6 Concluding Remarks

The industry is currently exploring cheap ways to mass-produce nanomaterials. Consequently, we will undoubtedly see more "natural" nanomaterials being used in commercial applications. Natural and modified natural nanomaterials would be good reference points for comparison of the functionality, cost, and potential ecological implications of synthetic nanomaterials. While natural nanomaterials are found in large quantity in the Earth, the process of extraction and separating them is technically challenging. Today, most natural nanocomposites used in the industry are very difficult and expensive to process.

For example, platy clays, such as the kaolin clay, contain layered two-dimensional lamellaes held together by an intercalation (or intermediary) layer. Thus, the lamellaes must first be exfoliated, or chemically separated, so they can function as nanofillers and be dispersed in the polymer matrix. In today's platy nanoclay production process, these multiple steps lead to complexity, cost, and dispersion quality challenges. In addition, most manufacturers do not have the specialized equipment needed to produce composites with platy nanoclays; this required investment is the biggest market barrier for nanoclays.

The same occurs for the use of nanocellulose-based materials since after a sequence of steps for the extraction of nanocellulose, a yield of around 30 wt.% of nanocellulose is achieved. The whiskers produced after the extraction treatment have great potential in nanobiocomposites applications.

As has been shown, many companies, especially in North- America, Europe and Asia, are commercializing not only processed/modified natural nanomaterials, but also additives and resins reinforced with nanostructured natural materials. This concern with eco-friendly materials is also observed in final users, like the examples cited in the automotive industry.

References

[1] Saheb DN, Jog JP (1999) Natural fiber polymer composites: a review. Advances in polymer technology 18:351-363

[2] Rajendran V (2009) Development of Nanomaterials from Natural Resources for Various Industrial Applications. Advanced Materials Research 67:71-76

[3] Guodong Y (2005) Natural and Modified Nanomaterials as Sorbents of Environmental Contaminants. Journal of Environmental Science and Health Part A 39:2661-2670

[4] Gao H, Ji B, Jäger IL, Arzt E, Fratzl P (2003) Materials become insensitive to flaws at nanoscale: Lessons from nature. Proceedings of the national academy of sciences of the united states of america 100:5597-5600

[5] Doktycz MJ, Simpson ML (2007) Nano-enabled synthetic biology. Molecular Systems Biology 3:125.1-125.10

[6] Grim PE (1968) Clay Mineralogy. 2nd ed. McGraw-Hill, New York

[7] Bergaya F, Theng BKG and Lagaly G (2006) Handbook of Clay Science. Elsevier Ltd., Amsterdam

[8] CIPEA (1996) Nomenclature Sub-Committee Minutes. Proc. Internat. Clay Conference, Israel

[9] Zhao C, Qin H, Gong F et al (2005) Mechanical, thermal and flammability properties of polyethylene/clay nanocomposites. Polym Degrad Stabil 87:183-189

[10] Ling Y, Hu Y, Lu H, Song L (2006) Morphology, thermal and mechanical properties of flame-retardant silicone rubber/montmorillonite nanocomposites. J Appl Polym Sci 99:3275-3280

[11] Gilman JW (2007) Flame retardant mechanism of polymer clay nanocomposites. In: Morgan AB, Wilkie CA (ed) Flame retardant polymer nanocomposites. John Wiley & Sons Inc., New Jersey

[12] Jang BN, Costache M, Wilkie CA (2005) The relationship between thermal degradation behavior of polymer and the fire retardancy of polymer/clay nanocomposites. Polymer 46:10678-10687

[13] Morgan AB (2006) Flame retarded polymer layered silicate nanocomposites: a review of commercial and open literature systems. Polym Adv Technol 17:206-217

[14] Laoutid F, Bonnaud L, Alexandre M et al (2009) New prospects in flame retardant polymer materials: from fundamentals to nanocomposites. Mater Sci Eng R 63: 100-125

[15] Kiliaris P, Papaspyrides CD (2010) Polymer/layered silicate (clay) nanocomposites: An overview of flame retardancy. Progr Polymer Sci 35:902-958
[16] Powell CE, Beall GW (2006) Physical properties of polymer/clay nanocomposites. Curr Opin Solid State Mater Sci 10:73-80
[17] Choudalakis G, Gotsis AD (2009) Permeability of polymer/clay nanocomposites: A review. Eur. Pol. J 45:967-984
[18] Meneghetti P, Qutubuddin S (2006) Synthesis, thermal properties and applications of polymer/clay nanocomposites. Thermochim Acta 442:74-77
[19] Paul DR, Robeson LM (2008) Polymer nanotechnology: Nanocomposite. Polymer 49:3187-3204
[20] Mai Y-W, Yu Z-Z (2001) Polymer Nanocomposites. Woodhead Publishing Limited, Sawston.
[21] Okada A, Usuki A (2006). Twenty Years of Polymer-Clay Nanocomposites. Mater. Eng. 291:1449-1476
[22] Choudalakis G, Gotsis AD (2009) Permeability of polymer/clay nanocomposites: A review. Eur Polim J 45:967-984
[23] Ray S S, Okamoto M (2003) Polymer/layered silicate nanocomposites: a review from preparation to processing. Prog. Polym. Sci. 28:1539-1641
[24] Usuki A, Hasegawa N, Kato M (2005) Polymer-Clay Nanocomposites. Adv Polym Sci 179: 135-195
[25] Chaiko DJ (2002) The Colloid Chemistry of Organoclays. In: SAMPE Conf Polymer Nanocomposites, Long Beach.
[26] Mirabella Jr FM (2004). Polypropylene and Thermoplastic Olefin Nanocomposites. In: Schwarz JA, Contescu CI, Putyera K (ed) Dekker Encyclopedia of Nanoscience and Nanotechnology. Taylor and Francis, London
[27] Choudalakis G, Gotsis AD (2009) Permeability of polymer/clay nanocomposites: A review. Eur Polim J 45:967–984
[28] Jancar J, Douglas JF, Starr FW et al (2010) Current issues in research on structure-property relationships in polymer nanocomposites. Polymer 51:3321–3343
[29] Pavlidou S, Papaspyrides CD (2008) A review on polymer–layered silicate nanocomposites. Progr Polymer Sci 33:1119-1198
[30] Tjong SC (2006) Structural and mechanical properties of polymer nanocomposites. Mater Sci Eng R Rep 53:73-197
[31] Alexandre M, Dubois P (2000) Polymer-layered silicate nanocomposites: preparation, properties and uses of a new class of materials. Mater Sci Eng 28:1-63
[32] Klemm D, Heublein B, Fink HP, Bohn A (2005) Cellulose: fascinating biopolymer and sustainable raw material. Angewandte Chemie-International Edition, 44:3358-3393
[33] Siró I, Plackett D (2010) Microfibrillated cellulose and new nanocomposite materials: a review. Cellulose 17:459-494
[34] Nakagaito AN, Yano H (2005) Novel high-strength biocomposites based on microfibrillated cellulose having nano-order-unit web-like network structure. Appl Phys A 80:155-159
[35] Nakagaito AN, Iwamoto S, Yano H (2005) Bacterial cellulose: the ultimate nano-scalar cellulose morphology for the production of high-strength composites. Applied Physics A - Materials Science & Processing, 80:93-97
[36] Zhang G, Zhang J, Xie G, Liu Z, Shao H (2006) Cicada Wings: A Stamp from Nature for Nanoimprint Lithography. Small 2:1440-1443
[37] Du ML, Guo BC, Jia DM (2010) Newly emerging applications of halloysite nanotubes: a review. Polymer International 59:574-582

[38] Cherny I, Gazit E (2008) Amyloids: Not Only Pathological Agents but Also Ordered Nanomaterials. Angew Chem Int Ed 47:4062-4069
[39] Klemm D, Schumann D, Kramer F, Heßler N, Koth D, Sultanova B (2009) Nanocellulose Materials — Different Cellulose, Different Functionality. Macromolecular Symposia 280:60-71
[40] http://www.vtt.fi/kuvat/uutta/Nanoselluloosakeskus_Qvintus_LeinoVTT.pdf Accessed in May 2010
[41] Iguchi M, Mitsuhashi S, Ichimura K, Nishi Y, Uryu M, Yamanaka S, Watanabe K Bacterial cellulose-containing molding material having high dynamic strength. USPTO 4742164, 1988
[42] http://www.sony.net/SonyInfo/News/Press_Archive/200404/04-0415E/ Accessed in May 2010
[43] Windebank M, Hart J, Alary JA. Proppants and anti-flowback additives made from sillimanite minerals, methods of manufacture, and methods of use. USPTO 7737091
[44] Floody MC, Theng BKG, Reyes P, Mora ML (2009) Natural nanoclays: applications and future trends — a Chilean perspective. Clay Minerals 44:161-176
[45] http://www.naturalnano.com/ Accessed in May 2010
[46] http://www.sigmaaldrich.com/materials-science/nanomaterials/nanoclay-building.html Accessed 15 May 2010
[47] http://www.nanocor.com/ Accessed in May 2010
[48] http://www.nanoclay.com/ Accessed in May 2010
[49] Christiani B, Maxfield M. Melt process formation of polymer nanocomposite of exfoliated layered material. USPTO 5747560
[50] Lee EC-C, Mielewski DF, Method for producing a well-exfoliated and dispersed polymer silicate nanocomposite by ultrasonication. USPTO 6828371
[51] Ross M, Kaizerman J. Clay/organic chemical compositions useful as additives to polymer, plastic and resin matrices to produce nanocomposites and nanocomposites containing such compositions. USPTO 6380295
[52] Ross M, Kaizerman J. Smectite clay/organic chemical/polymer compositions useful as nanocomposites. USPTO 6521690
[53] Ross M, Kaizerman J. Organoclay/polymer compositions with flame retardant properties. USPTO 6610770
[54] www.grupofiat.com.br/Biblioteca/UpLoad/MundoFiat/102.pdf Accessed in June 2010
[55] www.mercedes-benz.pt/content/portugal/mpc/mpc_portugal_website/ptng/home_mpc/passengercars/home/passenger_cars_world/environment_portugal/environments/value_chain/natural_fibre.html Accessed in March 2010
[56] www.revistasustentabilidade.com.br/reciclagem/fibras-naturais-e-sinteticas-sao-usadas-no-interior-de-veiculos-automotres Accessed in March 2010
[57] http://www.psa-peugeot-citroen.com/en/psa_group/innovation_b2.php Accessed in September 2010
[58] http://plasticsnews.com/headlines2.html?id=19202 Accessed in July 2010
[59] http://www.gm.com/corporate/responsibility/environment/recycling/index.jsp Accessed in March 2010
[60] http://www.mitsubishi-motors.com/corporate/environment/report/e/pdf/2008/2008e_07.pdf Accessed in March 2010
[61] http://www.toyota.com/about/environmentreport2009/03_recycling.html Accessed in March 2010

Abbreviations

BC - Bacterial cellulose
HDT - Heat distortion temperature
MFC - Microfibrillated cellulose
NMR - Nuclear magnetic resonance
PCN - Polymer-matrix clay reinforced nanocomposites

Author Index

GPSR Compliance

The European Union's (EU) General Product Safety Regulation (GPSR) is a set of rules that requires consumer products to be safe and our obligations to ensure this.

If you have any concerns about our products, you can contact us on ProductSafety@springernature.com

In case Publisher is established outside the EU, the EU authorized representative is:

Springer Nature Customer Service Center GmbH
Europaplatz 3
69115 Heidelberg, Germany

Batch number: 10584448

Printed by Printforce, the Netherlands